DESIGN, CONTROL, PREDICT

DESIGN, CONTROL, PREDICT

LOGISTICAL GOVERNANCE IN THE SMART CITY

Aaron Shapiro

UNIVERSITY OF MINNESOTA PRESS
MINNEAPOLIS • LONDON

Portions of chapter 2 were published in a different version as "Between Autonomy and Control: Strategies of Arbitrage in the 'On-Demand' Economy," *New Media and Society* 20, no. 8 (2018): 2954–71; copyright 2017 Aaron Shapiro, https://doi.org/10.1177/1461444817738236; and as "Dynamic Exploits: Calculative Asymmetries in the On-Demand Economy," *New Technology, Work, and Employment* 35, no. 2 (2020): 162–77. Portions of chapter 3 were published in a different version as "Predictive Policing for Reform? Indeterminacy and Intervention in Big Data Policing," *Surveillance and Society* 17, nos. 3–4 (2019): 456–72; copyright 2018 Aaron Shapiro, https://doi.org/10.24908/ss.v17i3/4.10410.

Published by the University of Minnesota Press
111 Third Avenue South, Suite 290
Minneapolis, MN 55401-2520
http://www.upress.umn.edu

ISBN 978-1-5179-0826-3 (hc)
ISBN 978-1-5179-0827-0 (pb)

Library of Congress record available at https://lccn.loc.gov/2020034655

Printed in the United States of America on acid-free paper

UMP BmB 2020

For Natalie and Nico

Contents

Introduction

Logistical Governance in the Smart City

Logistics is the art and science of managing the mobility of people and things to achieve economic, communication, and transport efficiencies.

—BRETT NEILSON, "Five Theses on Understanding Logistics as Power"

A key feature of smart cities is that they create efficiency.

—KENDRA L. SMITH, "The Cities of the Future Will Be Efficient, Sustainable, and Smart"

Technology is reconfiguring urban life. In the "smart city," data and information do more than just represent urban processes—they intervene in them. Data flows and information architectures structure our experience of urban infrastructure, mediating our access to institutions, resources, and services. Software and interface designs limit what information is available, how it can be retrieved, by whom and when. Algorithms translate data inputs into "actionable" outputs, anticipating future flows of people, materials, and information and interceding to guarantee or prevent certain outcomes. The technologies of the smart city calculate, calibrate, and mobilize; they identify and exploit increasingly minute sources of efficiency. And they are already being put to work, remaking our cities in the image of "urban intelligence"—transforming how public space is designed and administered, how work and labor are managed, how neighborhoods and communities are policed.

This book is about the smart city—not as some idyll techno-urban future toward which we are slowly but inexorably heading, but as a critical appraisal of urban governance today.

When the label of smartness attaches to technology (smartphones, smart watches, smart appliances, and so on), it calls to mind a particular set of sociotechnical affordances—connectivity, interactivity, autonomy. In the smart city, these affordances are imagined to extend to the social, material, and institutional realms of urban life: networked sensing devices in the built environment; digital dashboards for monitoring and administering city services; automated decision-making; ubiquitous connectivity; algorithms to anticipate future threats or new sources of profit. Turnkey smart city "solutions," brought to market as packages and accessible only through proprietary software, operationalize various combinations of those affordances. Silicon Valley giant Nvidia, well-known as the inventor of the graphics processing unit, promotes its Metropolis platform as "the foundation for smart cities." Metropolis leverages sophisticated artificial intelligence (AI)–based video analytics to identify inefficiencies across a range of city services—from retail analytics to traffic management, smart parking to law enforcement.[1] Former cellphone manufacturer Nokia's Integrated Operations Center works similarly. It provides "data feeds and alerts from various smart city devices, systems and applications"—such as street lighting and waste management—to "orchestrat[e] the 'smarts' in smart cities for efficiency, speed and new monetization opportunities."[2] Indeed, urban technologies of all flavors take advantage of some combination of these sociotechnical toolkits, even if not marketed explicitly as a "smart city solution." On-demand service platforms like Uber, for example, exploit workers' and customers' smartphone data plans as a connective infrastructure while marshaling complex algorithms to automatically adjust prices and pay rates according to statistical forecasts of supply and demand.

If investment is any indication, the smart city industry is on the rise. Between 2016 and 2018, venture capitalists invested about $76.8 billion in "urban tech" startups—about 17 percent of the VC funds invested globally during the same period.[3] Together, these companies promise to transform every dimension of urban life, from transportation and food delivery to co-living and co-working, construction to real estate.

According to market projections, the industry will be worth a combined $2.4 trillion by 2025.[4] On the public-sector side, municipal governments drive demand. City, state, and national governments are the primary "buyers" of smart city solutions, seeking out public–private partnerships for investment and management. In India, the national government subsidized massive urban renewal and technological retrofitting projects as part of its five-year plan to make one hundred cities smart by 2022.[5] Major partnerships between cities and corporations in North America herald the entrance of tech giants like Alphabet, Google's parent company, to the real estate industry. In the course of writing this introduction, Alphabet subsidiary Sidewalk Labs unveiled its master plan for Quayside, a twelve-acre "smart" district on Toronto's waterfront, promising nothing short of urban salvation through technological saturation—a city "built from the internet up."[6]

Together, these technologies, services, platforms, and developments furnish life in the smart city. While their trappings may seem disparate, I argue that the "smart city" coheres as a frame for intervention around a distinctively *logistical* set of concerns—optimization, rationalization, and coordination. These concerns impart a particular way of seeing and engaging the city, of managing urban life—capitalizing on its impulses while tamping out its excesses, all in service to the demands of "efficiency maximization."

Of course, municipal, state, and national governments have long sought to increase the efficiency by which urban flows of people, goods, and information are administered. But it is also the case that logistics finds new life in the smart city—operationalized by information technologies and honed by increased computational power. Indeed, the smart city's technologies lend themselves especially well to logistical concerns. When we talk about code and the city—about digitization, algorithms, AI, and the "internet of things"—we're talking about the very foundations of logistics.[7] As one computer programmer told me during the course of my research, "Nobody really figured out what to do with computers until the 1990s. That's when they figured out that what you could do is big-box chain stores, and it just completely wiped mom-and-pop businesses off the map—because they realized that what you could do is logistics."[8]

A similar paradigm shift is taking place in our cities. While most urban dwellers would probably agree that a little bit of efficiency could go a long way toward improving city life, the smart city takes efficiency to new extremes. As Brett Neilson defines it in his critique of logistics as power, logistics designates "the art and science of managing the mobility of people and things to achieve economic, communication, and transport efficiencies."[9] In the smart city, efficiency motivates intervention. It reconstructs urban worlds in the image of the supply chain—as an abstract, calculative surface; a smooth, infrastructural plane on which people, goods, and information must circulate with ever-greater efficiency. Coordination, rationalization, optimization: these imperatives no longer serve as the means to some broader or more utopian end; they are now core preoccupations of urban governance in and of themselves.[10]

At least according to the smart city fantasy. Efficiency ideals hardly reflect the reality of everyday urban life. Smart city interventions are fraught with contradictions, and when their efficiency demands encroach on new social and political registers, society's least advantaged feel the effects most directly. Platforms like Uber and vacation rental site Airbnb, for example, promised to democratize consumer access to luxury transportation and hospitality. In the cities where these services flourished, however, they have left behind trails of displacement, unemployment, and protest. In New Orleans, fair housing activists argue that the rise of short-term vacation rentals on Airbnb exacerbated the city's housing crisis. Just as the real estate market was rebounding from the long tail of devastation wrought by Hurricanes Katrina and Rita in 2005, landlords eager to cash in on the demand for short-term rentals started reserving their properties for Airbnb, decreasing the supply of units available for traditional twelve-month leases. This led to significant rent hikes, forcing some of the city's most economically vulnerable out to distant, less expensive suburbs, far from their work and families.[11] In New York City, Uber's entrance to the transportation market had a similarly dramatic impact on the taxi industry, which was already in the midst of a massive financial bubble. The value of taxi medallions, long required to operate a cab, plummeted from a peak of $1.3 million in 2013 (UberX's first year of operations) to as low as $160,000 in 2018. And because taxi workers' fiscal well-being and retirement had become so intertwined with the medallions' value, drivers were hit with an immense debt crisis associated with the suicide deaths of at least eight drivers in a single twelve-month span.[12]

These and other stories of technologies "disrupting" urban lives and livelihoods converge as a dark shadow over the rapidly growing market in urban tech. *Design, Control, Predict* studies the smart city's encroachments—the contradictions, frictions, and irrationalities of urban technological overhaul—to understand how logistical governance both guides and undermines our cities' futures.

A Constellation of Awkward Integrations

The phrase "smart city" is fraught with ambiguity and confusion—who or what is made "smart," how and why? Although the term has become commonplace among both practitioners and scholars as a buzzy shorthand for the types of sociotechnical affordances described above,[13] widespread usage has yet to foment consensus over what, precisely, makes a city "smart."[14] Competing definitions suggest different scales of intervention—from localized sensor networks and platforms to control rooms, dashboards, and indeed entire cities. It may even be the case that the only consistent feature of the smart city is its definitional uncertainty. According to critical technology scholars Jathan Sadowski and Roy Bendor, technology only becomes a *smart city* technology "by association with the idea of the smart city and the narratives, logics, practices, and symbolism of which it is constituted."[15] In other words, the smart city is as much a cultural construct as it is a technical one—a fact that underlines (rather than undermines) the need for critical empirical work.

As a result of its definitional fuzziness, critical researchers have approached the smart city from multiple angles, each providing insights into its cultural, political, and technical logics. First, with a growing number of technology giants entering the market (Cisco, Siemens, IBM, Qualcomm, Microsoft, and Alphabet, to name just a few), one stream of research analyzes corporate promotional materials—plans, prototypes, rhetoric, imagery, and videos—as a window onto corporate visions for the urban future. Geographers Ola Söderström, Till Paasche, and Francisco Klauser, for instance, found that the discourse of "smartness" in IBM's Smarter City campaign worked as a form of "corporate storytelling"—

a narrative that the company tells to establish itself as a necessary partner (if not driver) of urban development.[16] Sadowski and Bendor likewise argue that corporate rhetoric offers "a particular narrative about urban crises and technological salvation."[17] The smart city, when viewed through the lens of its boosters and their campaigns, is little more than corporate spin on a long tradition of technical solutions to "urban problems"—a fantasy of control that for-profit companies promote to win a seat at the lucrative table of urban governance.

A second approach focuses on a handful of canonical smart city developments—cities or districts saturated with sensing architectures and informatic infrastructures, built from scratch on green- or brownfield sites, and commonly held up as ideal-type models for smart city development. Examples include Songdo in South Korea, PlanIT Valley in Portugal, Masdar City in the United Arab Emirates, the Hudson Yards development in New York City, and, more recently, Quayside in Toronto. Each proceeds from a lucrative partnership between city, state, or national governments and technology companies, suggesting novel concentrations of corporate dominion over urban life.[18]

Songdo in South Korea, for example, has become particularly iconic of a new era of paternalistic corporate governance operationalized through ubiquitous computing.[19] Managed through a partnership between the South Korean government, Cisco Systems, and developer Gale International, Songdo is inundated with sensing devices and "doubly communicative" surfaces designed to convey information and gather data through the same interactions. Indeed, in Songdo, everything is datafied, from weather and environmental conditions to trash collection and pedestrian movements—all monitored and administered from centralized command-and-control rooms. According to media theorist Orit Halpern and her colleagues, Cisco's monopoly on Songdo's data transforms the entire city into a "test bed" for the commodification of new modes of urban life.[20] Under this data-intensive governance regime, the city becomes a space "for corporations to perfect the design of data collection and management infrastructures for any network—urban or otherwise."[21] We see a similar dynamic playing out (albeit with much more substantial grassroots resistance) in Toronto's Quayside development, as Alphabet tries to corner the market as both urban developer and systems administrator.[22]

In each of these cases, the distinction between the city as territory, community, and political body on one hand, and the city as a replicable and commodifiable model of technological life on the other, becomes nearly impossible to justify.

The third approach, which I build on throughout the book, studies what critical geographers Taylor Shelton, Matthew Zook, and Alan Wiig have called the "actually existing" smart city.[23] As a way of making sense of technology's impact on urban life today, the actually existing approach offers a more on-the-ground perspective. It prioritizes "situated understandings" of technological interventions, tracing where smart city policies come from and how they are implemented. Part of the rationale is suspicion of the magic-like "digital stardust" that animates corporate promotional materials and canonical developments like Songdo and Quayside.[24] And there's good reason to be suspicious. Gale International (the developer in Songdo) and Alphabet subsidiary Sidewalk Labs (Quayside's developer) both have a motive to exaggerate the smoothness of their operations. Gale is hardly likely to advertise that Songdo is having trouble attracting residents,[25] and Sidewalk Labs has barely commented on the departure of major government stakeholders over privacy concerns.[26] If we focus too intently on corporate plans or test beds for the urban future, we risk overlooking the on-the-ground conditions of cities today and how urban life has a tendency to frustrate smart city ambitions—which is exactly where the actually existing approach leads us.

Equally important are the growing number of smart city projects that do not involve massive urban development. Outside the skyscrapers and digital glitz of Songdo and Quayside, technological interventions are taking hold in "ordinary cities," where digital upgrades or "hacks" to legacy urban systems promise to revitalize infrastructure and services long neglected and underfunded after decades of economic decline and deindustrialization.[27] These retrofitting efforts expose cracks in the smart city façade—misalignments between the types of issues that information technologies can meaningfully address and the types of "urban problems" that technologists claim they solve. In many cases, interventions are prefaced by years of hype, only to be wildly disappointing.

Overhyped and underperforming. The actually existing smart city is Kansas City, where local and state governments in Kansas and Missouri waived millions of dollars in right-of-way fees for Google to lay fiber-optic cables for ultra-fast internet, only to have 70 percent of the region's school children still lacking home internet access after five years.[28] It is Louisville, Kentucky, where the city government undertook a massive effort to digitally map vacant properties, only to reveal already well-known disinvestments in African American neighborhoods.[29] And it is my hometown, Philadelphia, Pennsylvania, where the city government partnered with IBM to develop an "app-based solution" for jobs training at the same time that the public school district was being privatized and stripped of assets and funding.[30]

As these and a growing list of other examples show, technological fixes to urban problems do not happen in a vacuum, and the impurities matter. Smart city interventions are always "the outcomes of, and awkwardly integrated into, existing social and spatial constellations of urban governance and the built environment."[31] Those "awkward integrations" generate friction. There are learning curves and growing pains to be sure, as is expected when institutions integrate new tech into their existing procedures or partner with profit-seeking entities to fund urban infrastructure projects. But more fundamentally, new technologies *reconfigure social distances.*[32] Technology designs smuggle in assumptions about who makes up a system's user base and how they will use it, what the costs and benefits of the technology will be and for whom. As these assumptions rub up against existing practices and institutions, new and often unanticipated problems emerge—in some cases exacerbating the same disparities and inequities that the technology purported to solve.

Friction is thus a useful heuristic for understanding what the smart city looks like today, on the ground and in our communities.[33] It captures how "the rubber hits the road" in technology-based interventions while emphasizing just how uneven, maldistributed, and disputed the smart city can be as a lived experience. But as critical logistics researcher Kate Hepworth reminds us, merely pointing out the existence of frictions—highlighting the "seams in flows imagined to be seamless"—is insufficient.[34] Accounting for friction requires that we study the particularities and specificities that shape each "awkward integration" of technology in the city's built environment and institutions. Consider Nvidia's Metropolis and Nokia's Integrated Operations Center again. To understand how these "solutions" reconfigure services across sectors as disparate as retail analytics and waste management requires that we take stock of the social relationships, institutional norms, bureaucratic mechanisms, and material technologies that together constitute retail analytics or waste management as a coherent "problem space" to begin with; only then can we start to understand how new technologies affect operations. Each intervention comes with its own set of contingencies that, in turn, call for close, tailored attention.

And yet, smart city technologies are also remarkably flexible, adaptable, and mobile. To study the "actually existing smart city" does not mean *ignoring* connections across sites of intervention. Quite the contrary. Technological solutions travel—geographically, but also across problem spaces.[35] The very idea that Metropolis or the Integrated Operations Center could be useful for sectors as distinct as retail and waste management is a case in point. What sociotechnical logics connect and operate *across* the particularities of these domains—how are the department store and the dump made amenable to the same or similar technological interventions? If information is best understood as an "event within a system of organized memory" as critical theorist Suhail Malik suggests, then what implications will information technologies have for urban management—when the city is conceived as an interoperable "system of systems," not a space of difference but of informatic visibility and commensurability?[36]

Addressing these questions requires thinking across interventions while simultaneously attending to their localized frictions. My approach is thus to reconcile seemingly competing analytic objectives: to capture the specificity and friction of the actually existing smart city while engaging its awkward integrations as a constellation of broader transformations in urban governance—changes that, I will argue, are driven by increasingly *logistical* rationales.

Logistics is an art of synthesis. Where information technologies make flows and movements visible in particular ways, logistics tries to make them mutually intelligible—or "actionable," to use the parlance of practitioners. In the transnational networks of global supply chains, where logistical techniques and technologies were first honed and refined, logisticians integrated transportation with communication, distribution with manufacturing, consumption, and marketing.[37] The smart city adapts these synthetic intelligibilities to the management of *urban* activity and movement. Like the "lean" and "flexible" manufacturing and management ideologies that dominate global production circuits, logistical governance proceeds according to a "circulationist philosophy" of "continuous flow," where the rhythms of urban life are made to align through technologically enabled informatic visibilities.[38]

Logistics capitalizes on these syntheses by identifying increasingly minute sources of (in)efficiency across scales, from the territorial and economic to the cultural and corporeal.[39] The technologies of the smart city urbanize this same will to optimize and streamline. They reconfigure urban flows as an endless source of value or competitive advantage, because *flows can always be made more efficient*. If cities represent "the last great frontier of inefficiency in capitalism," as urban development guru Richard Florida submits,[40] then the smart city is about transforming those inefficiencies into a seemingly infinite resource, a raw material for capitalist value extraction.[41]

But logistics is also an agnostic art—it makes no commitments. Logistics designates the "means and arrangements" that enable and support movement and transmission, and it does so according to a highly variable, and ultimately political, criterion of efficiency. Those who operate the (computational) control levers quite literally define, tweak, and fine-tune the parameters of efficient movement—setting algorithmic weights, designing the interfaces, structuring the flow of information, determining which outcomes are "optimal." And while this agnosticism may serve strategic ends (in the sense that powerful actors can always define "efficiency" to mean whatever they would like it to), it also offers a glimmer of hope for the future of our cities. By approaching the actually existing smart city as a logistical project, my goal is to re-politicize the fetishization of efficiency, to examine the role that contingency, specificity, friction, and resistance play in a politics of efficiency and optimization. Whose efficiency? What is being optimized? Toward what end?

Means and Arrangements

Like strategy and tactics, logistics originates in military parlance. In the nineteenth-century treatise *The Art of War*, military strategist Antoine Henri de Jomini defined logistics as "the art of moving armies . . . the *means and arrangements* which work out the plans of strategy and tactics."[42] Embedded within this definition is a hierarchy that would define logistics for much of its intellectual history. In military operations, commercial enterprise, and even critical thought, logistics' coordinative rationalities were eclipsed by strategy's abstracting gravity and tactics' agile dexterity.[43]

This hierarchy would be upended by what observers call the "logistics revolution"—a global transformation in capitalism that tracked with technological development, deregulation, and the internationalization of commodity production and consumption.[44] Although logistical functions of coordination and distribution have always been essential to commerce,[45] the logistics revolution heralded a new prioritization of these concerns in management science and corporate practice. According to leading management scholar W. Bruce Allen, "Logistics is seen by its practitioners as the common link that weaves all the traditional functions of the firm together to meet customer requirements."[46] Indeed, for some of the world's largest multinational corporations operating today—Walmart and Amazon, for example—logistics informs strategy and tactics, not the other way around.[47]

The term "logistics" will likely conjure images of container ships and cargo ports, warehouses and last-mile distribution networks—the spaces and technologies of what anthropologist Anna Tsing has called supply chain capitalism.[48] And until recently, these logistical worlds received relatively little attention from social scientists and humanities scholars. The growing interdisciplinary field of "critical logistics studies" responds to that gap with an empirical critique of supply chain capitalism. Critical logistics research rejects the idea that logistics is apolitical. It calls attention to the power dynamics and violence that undergird global supply chain networks by emphasizing logistical regimes' "flaws, irrationalities, and vulnerabilities" and examining sites of struggle within logistical networks.[49]

In *Design, Control, Predict,* I absorb these critiques as I track logistics' mounting influence on the administration and management of digital urban networks. But my project also differs from much of this work. Rather than focusing on what we might imagine as the *urbanization of logistics* unfolding in the vast and growing network of "non-places" where global supply chain networks are territorialized in urban, regional, or extranational spaces (such as Dubai Logistics City or Shenzhen in China), my concern is instead with the *logisticization of urban life,* with the fetishization of efficiency and its reconfiguration of everyday life in the city. The conceit is to assume a more expansive role for critical logistics studies such that "logistics" designates not only an industrial field increasingly essential to the spatial and operational logics of capitalism, but also the growing influence of circulation and efficiency as a way of seeing, engaging, managing, and governing the city through the optimization of its flows.[50]

With this in mind, I use "logistics" throughout the book as a heuristic to capture a particular set of *problematizations* that motivate urban technological intervention. Historian and philosopher Michel Foucault spoke of "problematization" to describe how certain fields, sites, and practices are constructed as objects of thought and intervention. Problematization, Foucault explained, "does not mean the representation of a preexistent object nor the creation through discourse of an object that did not exist. It is the ensemble of discursive and non-discursive practices that make something enter into the play of true and false and constitute it as an object of thought (whether in the form of moral reflection, scientific knowledge, political analysis, etc.)."[51] The information technologies of the smart city operate on *the problematics of coordination and distribution,* where categories of "truth" and "falsehood" are weighed against "actionable" outcomes—measurable effects, optimal mobilizations, profits. If "the *actionable* is a task that can be executed with maximum efficiency," as critical logistics scholar Ned Rossiter suggests,[52] then what will be the consequences for a mode of urban governance that enlists "maximum efficiency" as a criterion of truth—or, indeed, as a condition for the kinds of urban problems that can be addressed in the first place?

The Logistics of Everyday Life

As a heuristic, then, logistics offers the most productive entry point to these questions. Specifically, logistics identifies three overlapping qualities of the smart city. First, following communications theorist John Durham Peters and others, the smart city activates through what we might call *logistical urban media*—technologies that facilitate, organize, and synchronize the city's movements, interactions, and transactions.[53] Like other logistical media, smart city technologies are primarily coordinative. They "arrange people and property into time and space." They have no content, per se, but operate on the basic coordinates of social experience—time and space. Logistical media are not messages; they activate "prior to and form the grid in which messages are sent."[54] They "calibrate labor and life, objects and atmospheres,"[55] structuring, arranging, and segmenting the rhythms and movements of urban life.

And indeed, some of humanity's oldest technologies are logistical, as are its urban media. Peters, for example, studies calendars, clocks, and towers—all technologies that have organized urban life for millennia.[56] While each medium may be more or less ornate, more or less ritualized in its form, its import lies not in the messages it sends or the symbols it bears, but in that it allows for messages to be sent in the first place. Logistical media coordinate and convene communication and sociality. Even in their antiquity, logistical media continue to have purchase—especially, Peters argues, after computerization: "New media have pushed the logistical role of media back to center stage."[57] Media scholar Liam Cole Young, for example, finds logistical media theory helpful for mapping the contours of "algorithmic culture," as "new spaces and times structure everything from the rhythms of life and labour to expectations regarding communication and commodity circulation."[58] Critical communications scholars Mark Andrejevic, Alison Hearn, and Helen Kennedy likewise turn to logistical media theory in their characterization of data mining.[59] Unlike traditional media, they argue, data mining does not involve narrative or representation; it is concerned with organization, indexing, segmentation, sorting, categorization—all logistical, not representational, functions, and all closely associated with the technological activation of "urban intelligence."

Second, logistics captures the instrumentality that the smart city imposes on city dwellers and urban sociality, or what political philosopher Julian Reid characterizes as *logistical life*—"life lived under the duress of the command to be efficient, to communicate one's purposes transparently in relation to others, to be positioned where one is required, to use time economically, to be able to move when and where one is told to."[60] Whereas Peters's logistical media arrange people and property in time and space, under the commands of logistical life smart city technologies interpellate people and property as *infrastructural relations*. Infrastructures are technological systems that enable or accelerate movement—networks that facilitate the flow and exchange of goods, people, or ideas.[61] In the smart city, technology arranges urban life into infrastructural networks, accommodating the kinds of movement that the administrators, managers, and operators of urban systems deem most productive or valuable.[62]

All infrastructures affect human culture and sociality in some way or form. Communications theorist James Carey famously showed that the complex interplay of telegraph and railroad networks transformed cultural understandings of temporality.[63] Meanwhile, to urban sociologist AbdouMaliq Simone, infrastructural relationality is a defining characteristic of cities. "Urbanization is not simply a *context* for the support or appropriation of specific lives," Simone writes, so much as "the plane upon which people—circling, touching, avoiding, attaching—come together, sometimes kicking and screaming, as an infrastructure."[64] Logistical life designates the instrumental appropriation of this plane of relationality, rendering the materials, institutions, and associations of urban-life-as-infrastructure legible, and therefore productive and governable. As Rossiter puts it, "Infrastructure makes worlds. Logistics governs them."[65] Like the "efficient individual" that critical theorist Herbert Marcuse identified with the technological society—"whose performance is an action only insofar as it is the proper reaction to the objective requirements of the apparatus"[66]—logistics referees what counts as action. We see its commands take hold on every front of the smart city's "control apparatus"—in the urban sensor networks that capture and register certain forms of activity but not others; in the way-finding smartphone apps that algorithmically carve out geographic trajectories based on aggregated smartphone location data; or in the facial recognition technologies that securitize transportation hubs and other sites of critical infrastructure.[67] These and other systems mimic what political geographer Anja Kanngieser observed among logistics workers—"a technological extension of governance onto the register of bodily movement and expression," the (technologically mediated) compulsion to communicate oneself, to make one's actions known, to be accounted for and submit to aggregation, to become infrastructure.[68]

Finally, as a problematization of efficiency and optimization, logistics designates the *technique* of mediation and interpellation—the means by which sensing devices, data, algorithms, "actionable" information, and so on integrate with the city's built environment and institutions.[69] According to philosopher Jacques Ellul, technique describes the relationships between technologies and the contexts of their application—how "the machine" is assimilated into the social order. "Technique has enough of the mechanical in its nature to enable it to cope with the machine, but it surpasses and transcends the machine because it remains in close touch with the human order." Technique "integrates the machine into society." It "constructs the kind of world the machine needs."[70] In the smart city, logistics constructs the kind of urban worlds that efficiency demands. "It clarifies, arranges, and rationalizes; it does in the domain of the abstract what the machine did in the domain of labor. It is efficient and brings efficiency to everything."[71] As technique, logistics closes the circle on mediation and infrastructural relations. It provides the most efficient rubric for efficiency maximization—and does so using the lightest touch possible, placing technologies "exactly where they ought to be and requir[ing] of them just what they ought to do."[72]

As a technique, then, logistics brings into focus the intricate reconfigurations required of technological integration and its instrumental remediation of urban times and spaces. But what efficiencies do those mediations prioritize? And with what frictions are they met?

Situating the Smart City

As we will see throughout the book, advocates of the smart city insist on the inherently innovative or disruptive character of new urban technologies.[73] In a 2019 special issue of the journal *Technological Forecasting and Social Change*, for example, management specialists and policy experts define smart cities as those that nurture "innovation ecosystems." Computer scientists promote "living labs" and "information marketplaces" as the foundations for smart city "innovation districts." Spatial data scientists insist that city governments must deregulate to allow for disruption, with little regard for the impact on communities.[74] Not many in industry are likely to refute these narratives—business depends on it.

This emphasis on the new and novel obscures the ample historical precedents to smart city interventions.[75] For one, modern political authority has always been information intensive. The early modern states in Europe consolidated their dominion with the information they gathered on population and territory.[76] Spatial science likewise has a long history that the "new" science of urban informatics rarely acknowledges.[77] Criminology dates to nineteenth-century France, with sociologists applying statistical techniques to crime data to study the correlation between location and "social character."[78] Chicago School sociologists revived this correlation in the 1920s and '30s to analyze urban geographies as determinants of social outcomes.[79] Meanwhile, private interests have long invested in "big data" in efforts to expand mass consumption and finance, with creditors and marketers using consumers' residence as a proxy for their economic traits, such as purchasing power and credit worthiness.[80] Nor are computational responses to "urban problems" especially new. Government-sponsored redlining maps in U.S. cities carved out spaces of disinvestment and exclusion for minority communities; and with computing available after World War II, RAND and other Cold War institutions applied insights from cybernetic warfare strategies to address the social problems proliferating in those same disinvested communities.[81] In the 1980s, researchers and computer scientists developed AI for urban planning decisions, such as zoning and land use policy;[82] and by the '90s, information and communication technologies threatened to upend traditional geographies of work, leisure, and commerce as engineers worked to bring about a new era of ubiquitous computing and scholars debated the rise of "cybercities."[83]

In these histories, it's possible to trace the technical and epistemological foundations of the smart city's "intelligence" back hundreds of years—to the rise of the modern state, capitalist political economy, and its techno-sciences.[84] The smart city, viewed from this wide angle, becomes just the latest in a long string of informatic and technological reconfigurations of the city as an object of thought and intervention.[85] Insisting on the historicity of urban technology and its relationship to urban governance pushes back against industry narratives of innovation and disruption. How and why did the techno-logics of the smart city come to cohere around concerns of circulation, mobility, organization, and transmission?

What motivates these interventions, where did the tools come from, and why now?

Austerity Urbanism

Today, the smart city connotes urban technology. But this was not always the case. Scholars emphasize that "smart city" has meant different things to different actors over time, from early designations of environmentally friendly land use allocations to policies for attracting and retaining skilled workforces.[86] In fact, the phrase only seems to have consolidated its close association with technology in the late 2000s, when large technology firms began transitioning from the heavy industry of hardware production to software- and service-based consulting. IBM's Smarter Cities campaign, for one, branded the company's pivot from legacy computer manufacturer to expert steward of technical knowledge. As geographer Donald McNeill shows, the mainstreaming of smart city discourse and its connection to advanced tech cannot be divorced from the efforts of multinational firms to cultivate a market for the more agile industry of "value-added" software and consulting services.[87]

By the time this industrial transition was underway, neoliberal economic policies had been chipping away at public infrastructures for decades—a process that urban geographers Stephen Graham and Simon Marvin described in 2001 as "splintering urbanism."[88] After the 2008 global financial crisis, however, neoliberalism's privatization of public goods and services only intensified, with governments imposing regimes of economic discipline by extreme austerity.[89]

Austerity is a defining feature of neoliberal economic thought and governance. In the United States and Europe, austerity's influence can be traced to the 1970s and '80s, with states translating the pro-market, anti-government conjectures of Chicago School economists into domestic policy.[90] But the mechanisms that mark austerity at its most extreme were first refined in the postcolonial states of the "developing world," where loans from the International Monetary Fund and World Bank had been used since the 1950s and '60s to effect dramatic economic reforms known as structural adjustments.[91] These not only constricted public-sector spending in postcolonial states already plagued by rampant and often colonially authorized poverty, and thereby ballooning corporate profits and exacerbating national (and global) inequalities; they also imported neoliberal ideologies of market-based solutions that opened borders up to mobile financial capital and offshore manufacturing operations. The result was a vicious cycle. As neoliberal reforms and structural adjustments aggravated national debt crises, budget deficits justified the further gutting of public expenditures and compelled even more privatization, resulting in an enforcement of market logics by fiat.

After the 2008 global financial crisis, this extreme austerity came home to roost in the metropoles of the Global North.[92] North American and European governments, at the municipal, state, and national levels, faced massive budget cuts. Dramatic spikes in unemployment ate into income taxes while the colossal epidemic of home foreclosures driven by the mortgage crisis depleted property-tax receipts. These trends were only exacerbated by several knock-on conditions: government bailouts of failing banks, for example, made public stimulus packages a political nonstarter, while corporations' increasing reliance on tax havens and offshore accounts to house their reserves helped them avoid taxation, further curbing government revenue.[93] As a result, just as millions across the globe were losing their jobs and homes, governments were cutting social spending. And although the crisis was triggered by an over-accumulation of fictitious capital in the private sector—and decidedly not governmental frivolity—the resounding response was to tighten the belt.[94]

As critical geographer Jamie Peck and others have shown, the post-2008 wave of austerity interventions involved more than a mere revival of neoliberal privatizations.[95] In the wake of crisis, governments used fiscal policy to intensify and consolidate advanced capitalist logics. The effects were "politically, socially, institutionally and fiscally cumulative."[96] Cities—already a favorite target of neoliberal fiscal policy interventions—bore the brunt of these austerity burdens. Although long recognized as engines of economic activity,[97] cities localize societal inequalities in geographic concentrations of social need and marginalization—what Karl Marx called the "hospital" of the working class.[98] To elites, such concentrations are rarely understood as symptoms of broader social or economic injustices; rather, they are the expected by-products of the inherently troublesome and unruly nature of cities. Since industrialization, conservatives have viewed cities as hotbeds of vice, poverty, and congestion, wellsprings of societal immorality and impropriety, inciting both intensive urban policing and vast reform campaigns.[99] Lacking in federal or state funding, municipal governments were forced to create social safety nets for themselves, investing in local social services and programs when state or national governments would not.[100] Conservative politicians continue to view such investments with disdain, as symbols of bloated bureaucracy, welfare, and public-sector spending, and therefore ideal targets of austerity reforms.[101]

After 2008, municipal budgets were slashed. In severe cases like Detroit, where the outward flow of people and capital had begun decades earlier, austerity urbanism exacerbated the crisis, with the city government forced to declare bankruptcy.[102] Even in less extreme situations, city services and public-sector benefits were cut dramatically. Municipal agencies were being asked to "do more with less"—a common refrain, which meant continuing to police, shelter the homeless, or even simply issue licenses and permits, only now with significantly smaller staffs and without traditional incentives like overtime pay. Much-needed and long-awaited infrastructural upkeep and repairs were postponed. Regulatory oversight waned in many sectors. Education budgets were ransacked as entire school systems were privatized in a fiscal variation of what Naomi Klein has called disaster capitalism.[103] As economist Paul Krugman wrote for the *New York Times*, the austerity drive "isn't really about debt and deficits at all; it's about using deficit panic as an excuse to dismantle social programs."[104]

It was in this political climate that smart city discourse began circulating as a technological solution to failing urban services. With tech companies like IBM and Cisco in the process of "streamlining" their business models, city governments and municipal agencies provided the ideal market for their new focus on value-added consulting and software services. In some places, the smart city was explicitly pitted as an austerity measure. Geographer Andrea Pollio found in Italy that the "smart city" served as a "political device" to "frame a number of new national and local policies as 'technical' solutions to low budgets and economic stagnation."[105] Burgeoning smart city tactics built on older neoliberal urban strategies, for instance, by trading massive tax benefits for corporate investment despite already-austere budget cuts, now merely under the guise of technology. The dynamics played out similarly in other contexts, even when the connections to austerity were less explicit. In Great Recession–era Philadelphia, for example, Alan Wiig charted discrepancies between the "smart city policy script" of technological solutionism and the actual benefits of technology-driven initiatives, nearly all of which were monopolized by the city's elites while social services continued to be stripped.[106]

The smart city emerges in mainstream discourse and political rhetoric as a set of technical fixes to what is ultimately the ongoing crisis in the welfare of society's most vulnerable—an upward redistribution of resources from public agencies and institutions to corporations, their investors, and their shareholders. Sold as productivity amplifiers and efficiency "hacks," smart city platforms promised to get city services back online, now with smaller staffs, slashed budgets, and fewer assets, all thanks to the assistance of pay-to-access platforms or dashboards. And as a project of crisis management, the smart city had the perverse effect of naturalizing even the most extreme and unnecessary austerity measures as the "new normal," just as the actual drivers of the financial crisis—disinvestment, privatization, economic exclusion—continued to propagate.

Flexible Accumulation

If post-crisis austerity contextualizes the governmentality of smart city solutionism, then the precise mechanisms of its interventions trace to deeper political–economic shifts unfolding throughout the second half of the twentieth century. As the IMF and World Bank leveraged their fiscal authority to impose neoliberal reforms around the globe, they also opened up new markets for labor and consumption. The internationalization of capitalist production had massively disruptive economic effects in the "developing" world, where it wrought often traumatic, even violent societal transformations.[107] But internationalization also affected the wealthier markets of the Global North by upending a delicate Keynesian bargain that had been struck between manufacturers and labor.[108] When commodity producers began offshoring manufacturing operations to markets with cheaper labor costs (often the same countries affected by austere structural adjustments), they could then turn around and sell their products for less than their competitors in domestic markets. This had cascading effects, not only in terms of wage stagnation; it also introduced fierce price competition that overwhelmed the Fordist economies. More commodities were being produced at lower prices than companies could sell, leading to overlapping crises of over-production, over-accumulation, and devaluation that wracked the developed world through the 1970s.[109]

These crises prompted commodity producers to rethink fundamentals, culminating in wide-ranging shifts in capitalist production—from the assembly-line manufacturing of Fordist mass production to the more flexible mode of accumulation that would define the post-Fordist political economy. The transformation was rooted in two reconfigurations, one technological and the other organizational. The most basic was technological. Under Fordism, the high, fixed costs of manufacturing equipment (or, in Marx's terms, the high cost of constant capital) made the threat of over-accumulation perilous. Machinery designed for the assembly-line was both expensive and rigid in its specialization, and therefore difficult to adapt to the demands of heightened competition or shifting consumer preferences. According to political economist and geographer Erica Schoenberger, because of this rigidity the Fordist economies of the 1950s and '60s were made up of relatively stable and homogeneous industries, with products differentiated by only superficial qualities.[110]

Internationalization upset this order by pressuring firms to diversify and seek out ever more granular sources of competitive advantage. And it was new technologies that facilitated this diversification and competition. As Schoenberger writes, "The key distinction between flexible technologies and traditional mass-production techniques is that what the machine actually *does* is programmed in via computer software rather than built into the machinery at the outset."[111] In hindsight, this upgrade seems straightforward, but it prefaced much broader social and organizational changes. Data and information became indispensable to workflow management and administration. As manufacturing and other workplace procedures were automated or semi-automated through computerization, managers gained access to a trove of data that could be used to streamline operations. Observing these changes throughout the 1980s, organizational sociologist Shoshana Zuboff argued that what made the new production regime truly different was not only that it automated production, but that it *informated* the process, reflexively generating data about machinic operations and work flows. "The devices that automate by translating information into action also register data about those automated activities, thus generating new streams of information."[112] And with computerization increasingly the norm across industries, the surfeit of information on work and production ushered in a new era of Taylorist management, marked by an obsession with productivity, flexibility, and efficiency.[113]

The second reconfiguration was thus organizational. With computer simulations and modeling, the Taylorist impulse to increase productivity by maximizing efficiency could now be extended throughout a commodity's life cycle—from material sourcing to assembly, distribution, and consumption. As equipment became more flexible in output—both quantitatively in terms of production speed and qualitatively in terms of the diversity of offerings—so, too, did the manufacturing process.[114] Whereas the Fordist firm oversaw the entirety of operations through vast communications and transportation infrastructures,[115] the "vertically integrated" management of mass production likewise came to be seen as overly rigid and expensive in the face of internationalization and over-accumulation. New and more "flexible" organizational schemes promised to overhaul the managerial end of production, making the workflow "leaner," more "agile," more responsive to uncertain market conditions—and certainly cheaper.[116] The result was what economist Richard Langlois has described as corporate *de-verticalization*, where previously highly integrated firms *dis-integrated* by spreading the production cycle across dispersed sites of activity, all configured in buyer–supplier relationships.[117]

Neither flexible accumulation nor economic globalization are possible without some degree of organizational dis-integration. Offshoring, by definition, requires dis-integrating the production process: firms in wealthy countries outsource manufacturing functions to contractors in markets where labor is cheaper. But dis-integrating also introduces coordination challenges that the traditional, vertically integrated firm avoided. Because independent suppliers of, say, machine parts are not the direct subordinates of an appliance manufacturer, the latter cannot oversee or directly manage the former's operations. The manufacturer loses control over "upstream" output. De-verticalization therefore created a demand for new and more sophisticated tools of coordination, and computerization again came to the rescue. Information and communications technologies would now ensure efficiency and consistency across vast distances (a managerial philosophy often called "total quality control") by monitoring, tracking, and informating—not only the production cycle, but the *distribution infrastructures* that this now required, including the movements of workers and inventory along each step of the supply chain. Informating technologies generated the data, computer models and simulations identified new (in)efficiencies, and firms began "to compete on the basis of the distribution of goods and services rather than merely the products themselves"—precisely what we now characterize as supply chain capitalism.[118]

The technical and organizational prioritization of flexibility ignited the logistics revolution mentioned above.[119] According to critical logistics scholar Martin Danyluk, the explosion in logistical activity beginning in the 1970s signaled nothing less than a global transformation in the capitalist system—a "logistical fix" to the crisis of over-accumulation.[120] If offshoring and outsourcing reconfigured operations on the factory floor, then logistics would be the counterpart on the distribution side—and computerization was equally important. Since the 1960s and '70s, logistics has inspired increasingly small and ever-more-powerful computational technologies and infrastructures. In 1962, supply chain guru Peter Drucker described logistics as "the economy's dark continent," likening distributive efficiency to the racialized frontier but lamenting that the technology of the day was ill suited to the task of colonization. "Only the high-speed computer," Drucker feared, "can analyze anything as complex as the distributive system."[121] Since then, logisticians have responded by pioneering a host of communications and transportation technologies and information infrastructures—bar codes, radio-frequency identification tags, drones, remote-controlled trains, flexible spreadsheets, map-based interfaces, enterprise resource planning systems, and so on—to illuminate the vast frontier that logistics represents.[122]

With these new tools in hand, logistics became the technical and organizational backbone of flexible accumulation. While the impact may have been global, the effects were felt locally. In American cities under

Fordism, the economy could guarantee financial stability and even economic mobility for working-class populations—at least the white working-class majority.[123] Geographically, this translated to relatively stable urban social topographies, with inner-city working-class white-ethnic enclaves and moderate middle-class (white) suburban growth. By the 1960s, however—after waves of African Americans fleeing state-sanctioned terror in the U.S. South resettled in manufacturing districts in northern cities—members of working-class white and ethnic white communities, as well as the established Black middle classes, began a process of resegregation, immigrating to the fast-growing and federally subsidized suburbs.[124] For those who remained, the disinvestment of white and middle-class flight was compounded by the flight of manufacturing jobs pushed offshore during the shift to post-Fordist production regimes; the factory jobs that did remain became increasingly scarce and wage-suppressed, undermining the mobility—and now even the mere stability—of urban working-class employment.

Flexible accumulation therefore also affected the racial and classed composition of North American cities.[125] The resulting formation is what geographers now call the "post-Fordist city." No longer a hub of production, the post-Fordist city excels in consumption-oriented services for the wealthy elite (who continue to live in suburbs or revitalized and securitized downtown enclaves) while concentrating racially and economically marginalized communities into increasingly isolated zones of exclusion.

The smart city *is* the post-Fordist city, both in that its geographies reinforce the uneven political–economic landscapes of flexible accumulation, and in that it integrates the same technologies designed to accommodate flexible manufacturing, only now in the city's infrastructures and built environments—not only to coordinate supply chain flows, but to monitor, track, and trace *urban* flows, of people, goods, and information.[126] We see these logistical logics on full display, for example, when the U.S. Department of Transportation's "Smart City Challenge" invites cities to compete on "intelligent" infrastructures for urban logistics and freight,[127] while Indianapolis mayor turned smart city guru Stephen Goldsmith implores cities to upgrade their governing "operating systems" to run more like Amazon—the logistics corporation par excellence.[128]

Phenetic Surveillance

As the political–economic crises of internationalization and over-accumulation spurred a revolution in logistics, its techno-logics of efficiency, optimization, and flexibility now offered a toolkit for managing urban life in an era of extreme austerity. But it was the post-9/11 explosion in surveillance that most immediately kickstarted the smart city's logisticization of urban life.[129] As any student of surveillance can attest, the proliferation of surveillance technologies in the first decades of the twenty-first century was an overdetermined phenomenon—it can be explained by any number of overlapping developments, including (but not limited to) (1) the securitization of airports, borders, and "critical infrastructures" following the 9/11 attacks; (2) the militarization of domestic policing and, in parallel, the civilianization and commoditization of what were previously military information infrastructures (such as the internet and GPS); and (3) the emergence of a mode of capitalist accumulation centered on affect, computer-mediated attention, and anticipation.[130] Disentangling these developments has spawned entire academic journals and edited volumes. Here, it should suffice to say that at each juncture, surveillance activates at the nexus of informatic visibility and mobility.[131] And it is at this same nexus that logistical concerns come to dominate the smart city's "urban problematic."[132]

According to sociologist and surveillance scholar David Lyon, surveillance operationalizes a classificatory impulse—what he calls a "phenetic fix," the drive "to capture personal data triggered by human bodies and to use these abstractions to place people in new social classes of income, attributes, habits, preferences, or offences, in order to influence, manage, or control them."[133] This social sorting serves a hinge function for the post-Fordist city's Janus-faced governance, with one visage presiding over mounting exclusion, segregation, and disinvestment, the other over an eruption of opportunity for privileged social groups. Surveillance's phenetic impulse provides the machinery for what sociologist Loïc Wacquant describes as the post-Fordist state's "Centaur-like" governmentality—"uplifting and 'liberating' at the top [of the class structure], where it acts to leverage the resources and expand the life options of the holders of economic and cultural capital; but . . . castigatory and restrictive at the bottom, when it comes to managing the populations destabilised by the deepening of inequality and the diffusion of work insecurity."[134]

Social sorting makes this bifurcated social governance possible, and it is felt especially acutely in the post-Fordist city, the city of spectacle and security, with "people and objects . . . mobilised, monitored and filtered between fortified places."[135] According to urban theorist Peter Marcuse, the post-Fordist city concentrates in two distinct formations—*the citadel* and *the outcast ghetto*.[136] Borne of forces closely associated with austerity governance and flexible accumulation—technological advances, internationalization, the privatization of public goods—the citadel and the outcast ghetto enact hyper-segregation as a defining feature of urban life. The citadel designates "spatially concentrated area[s] in which members of a particular population group, defined by its position of superiority, in power, wealth, or status . . . congregate as a means of protecting or enhancing that position."[137] Its boundaries are both symbolic and material, fortified with walls, gates, security guards, and surveillance infrastructures, all designed to sort out and exclude unwanted populations. The outcast ghetto, by contrast, houses what in Marx's terminology would be called the "lumpen proletariat"—a racialized economic-surplus population.[138] Unlike the "classic" ghetto, where racial or ethnic-minority residents were isolated but nonetheless able to participate in the mainstream economy, the outcast ghetto is determined as much by economic abandonment as it is by social isolation. Its boundaries are equally symbolic and material, designed not to protect social power but to limit it through containment.

The citadel and the outcast ghetto are to Marcuse inherently spatial formations—"concentrated areas"—and as such, defined by fixity. We continue to find these prototypic formations reproduced in certain "smart" urban revitalization efforts—for instance, in Camden, New Jersey, where a nearly $2 billion investment in waterfront development was fortified both symbolically and materially by a highly centralized, citywide surveillance network. In this and similar revitalization efforts, the investment zone follows "from a surveillance-first community policing strategy that would secure the city in advance of new enterprise and then, potentially, new residents."[139]

But there is also sense in which this spatial fixity is deceptive; that the citadel and the outcast ghetto describe not places per se, but rather unstable categories of personhood—racialized and classed valuations of populations' worth and risk, value and danger. Urban populations are hardly "immobile and frozen," and it is their mobility and flow—rather than their fixity and embeddedness—that motivates contemporary surveillance and security regimes.[140] Mobility is the essential condition of logistics. But in the post-Fordist city, mobility can become a source of anxiety to city leaders and corporate managers as social and economic "outcasts" circulate in menacing proximity to the securitized citadel. Technologies of separation and exclusion—the wall, the border, the cell, and the checkpoint—are themselves constantly on the move, monitoring, tracking, tracing, demarcating, excluding, and reconstituting urban spaces as matrices and abstractions of security and danger.[141] So, too, are the accoutrements of the citadel, where resources are invested to expand the amenities available to privileged social groups through a growing number of mobile services and provisions.

Surveillance's phenetic drive mediates the smart city's fractured and uneven mobilizations. As science fiction writer and technology theorist Bruce Sterling warns in his critique of the smart city, "The 'bad part of town' will be full of algorithms that shuffle you straight from high-school detention into the prison system. The rich part of town will get mirror-glassed limos that breeze through the smart red lights to seamlessly deliver the aristocracy from curb into penthouse."[142] What Sterling describes here is the extreme of a bifurcated logistical regime: the logistics of abandonment and the logistics of leisure and pleasure. Catering to the elites of the (mobile) citadel, logistics synchronizes circulation, ensuring that the movement of bodies and goods, services and information, proceeds as smoothly and efficiently as possible.[143] In the (mobile) outcast ghetto, logistics orchestrates the control and management of surplus populations, keeping them in their (social and economic) place, even as they move about the city.[144]

These logistically enabled, mobile segregations resonate with broader trends toward bifurcation and contradiction. Social theorist and geographer Nigel Thrift, for example, describes the increasingly conterminous application of surveillant control and mobile service provision as a "security-entertainment complex," a perpetual mediation between "permanent and pervasive war and permanent and pervasive entertainment."[145]

Media scholar Sarah Sharma similarly describes a post-9/11 prolifera-tion of non-places, where humans are at once reduced to "bare life" and invested in, as "lifestyle"—where the theme park and the refugee camp overlap as twin *nomos* of biopolitical authority.[146] These and other ampli-fied segregations force us to reckon with the smart city as a Centaur city, with information technologies as an investment in both opportu-nity and control, its logistics an intensification of already-uneven resource allocations.

Interventions

Austerity governance, flexible accumulation, phenetic surveillance—these social, economic, political, and technological forces prefigure the smart city as it actually exists today: as the uneven enactment of a mode of governance driven by the logistical imperatives of extraction and opti-mization, opportunity and control.[147] The critical task is to investigate how these dynamics play out, on the ground, in our cities and communi-ties today.

The book is organized around three case studies, each tailored to a distinct techno-logic of the smart city's logistical governance—design, control, and prediction. Chapter 1 considers the politics of technology design for public spaces by studying the design history of LinkNYC, a network of high-speed Wi-Fi kiosks replacing New York City's public payphones. LinkNYC is notable for two reasons. First, tech insiders and consultants point to LinkNYC as a model for public–private partner-ships to finance smart city infrastructure.[148] Second, LinkNYC is now the largest digital out-of-home urban advertising network in the world.[149] This "intersection" of marketing and infrastructure is not lost on Inter-section, the company that manages and operates LinkNYC and which describes its offerings as "Outdoor Advertising | Smart City Solutions." With Wi-Fi kiosks that double as digital billboards, Intersection marries messaging to sensing, advertising to surveillance. It translates the pro-grammatic interactivity and engagement of online adtech into the physical spaces of our cities, streets, parks, and plazas—another "intersection" of sorts. In doing so, LinkNYC realizes a decades-old dream of cybernetic transparency in the economy of ad circulation.[150] Through the network's "value-added" analytics of audience targeting and attribution metrics,

LinkNYC heralds a distinctly logistical future for both advertising and our cities' public spaces.

But LinkNYC's development also followed a fraught path. The chapter draws on publicly available documents, reporting, and interviews with designers to trace the network's trajectory from public design competition to procurement and implementation, highlighting key moments of friction along the way and revealing the political work needed to shepherd smart city interventions to fruition. There is the work of securing franchise bids, but there is also the cultural–political work of consolidating anticipation, of articulating a recognizable "semiotics of futurity" around designs to certify a sense of inevitability for a particular set of features—in this case, surveillance-driven advertising. Against claims from New York City Mayor Bill de Blasio that these features are necessary for bridging the city's digital divide, I argue that it was advertising's logistical demands—and not public access to information—that had the greatest impact on the network's design, and that this influence has hobbled our ability to imagine a more progressive and democratic alternative.

Interfaces remain a key motif through chapters 2 and 3, both of which study data infrastructures for managing mobile workforces. Chapter 2 focuses on workers in the platform or "on-demand" economy—the logistical underbelly of economic life in the smart city. Whereas proponents celebrate firms like Uber for ushering in a new era of efficiency, accessibility, variety, and convenience, the outlook is much grimmer for the platform workforce. Sold on the promise of flexible scheduling and workplace autonomy, platform workers struggle to navigate the automated management and information asymmetries of obscure policies and opaque algorithms. But I also show that platform workers' experiences cannot be boiled down to technical advances in algorithms or smartphone apps. Technology may serve the platforms as helpful subterfuge, but what on-demand companies truly thrive on is regulatory arbitrage—the misclassification of workers as independent contractors, not employees.

Technology enters the picture as an integrated tool of managerial control—the means and arrangements by which platforms administer their "flexible" workforces without violating workers' protections as independent contractors. I explore these dynamics through interviews with platform workers and through my own ethnographic observations, collected while working as a courier for the food-delivery startup Caviar. Whereas the platform economy is celebrated for its flexibility, I find that ultimately it is the firms, not the workers, that reap the lion's share of flexibility's benefits—the "flexibility" to change policies or procedures without accountability, to update the interface and remove vital information, to bring on more workers than can reasonably be paid and at increasingly low wages. Workers, for their part, are left to negotiate between the platforms' extractive calculus, their own well-being and sense of dignity, and the pragmatic challenges of navigating the city's uneven geographies of consumption.

Chapter 3 continues on the theme of automated management but shifts to workers of a wholly different sort—police officers. The chapter studies the production and design of a "predictive policing" platform, which uses machine learning algorithms trained on historical crime data and other information to predict where and when crimes will be committed and thereby allocate patrol resources more efficiently. At a moment when law enforcement agencies are struggling to maintain legitimacy, predictive policing has been sharply criticized by civil rights advocates as an expansion of police power and an entrenchment discriminatory policing practices. Meanwhile, the technology's advocates insist that predictive policing is about making policing fairer, less biased, more equitable—taking discretion away from line officers by rationalizing patrol routes and routines. The chapter navigates this debate by investigating system designers' perspectives, who believe that both claims may be true at once.

The chapter builds on ethnographic research with the product team at HunchLab, a geospatial predictive policing software produced by Philadelphia-based software analytics firm Azavea. Azavea's commitments to social justice made HunchLab unique within the police technology sector. It also made HunchLab a unique site to explore the ethical frontiers of smart city interventions. The company's sensitivity to technology's role in police reform forced me to reckon with crime prediction in a longer lineage. I explore the tensions and anxieties that new law enforcement technologies have always aroused in attempts to improve the relationship between the police and the polis. To fully understand predictive policing, I argue, we must engage with the institution of policing itself—an institution that has always been defined by the police's ambiguous role as protector of the peace with a license to inflict harm on the population. In the end, I conclude that technical solutions like HunchLab will never resolve policing's contradictory logistics as the distributor of both public safety benefits and state violence. Without dismantling police institutions, the police's relationship to society will always struggle with these competing mandates—the algorithm cannot save the institution.

What do these stories of logistical frictions tell us about the "smart city," about the technological reconfiguration of urban life? The book concludes with a chapter devoted to thinking through the fundamental problem that the previous chapters share: calculative mobilizations as the contradictory enactment of political, economic, and social power. The fantasy of the city as a smooth infrastructural surface—as a realization of cybernetic transparency, of perfect alignments between data and urban life, information and governance, supply and demand—is enormously compelling. But the frictions and resistances that recur throughout the pages of *Design, Control, Predict* illustrate exactly how illusory the dream is. The smart city has winners and losers, and by allowing the fantasy to overwhelm the analysis, we risk overlooking the extent to which the distribution of the smart city's costs and benefits is itself a political artifact. I therefore end by theorizing a "counter-logistics" of community care and support, both as a way to make sense of logistics as power and to begin imagining logistical applications to support a more equitable distribution of urban resources and opportunities.[151] How could a counter-logistical project help fortify our communities through meaningful and non-commercially mediated connection—to create spaces of repair and mutual protection?[152] We cannot answer this question if we think only of the kind of city that we want to build for the future. We need to interrogate the kinds of cities we are already living in—to investigate what can be done to make our existing institutions, governing bodies, and infrastructures more hospitable, ethical, and caring—not simply more efficient.

The City, in Transit(ion)

The idea for this book started in transit. As a graduate student, my daily commute took me from my home among the densely packed rowhomes of South Philadelphia's Point Breeze neighborhood across the Schuylkill River to the leafy campus of the University of Pennsylvania. Like Germantown Avenue for sociologist Elijah Anderson in *Code of the Street*, my rides across those neighborhoods gave me access to "an excellent cross section of the social ecology of a major American city."[153] I studied this cross section almost every day, observing the changing built environment and social landscape of my hometown. And there is much to say, even just about the twenty or so blocks between my old house and campus. I rode through Graduate Hospital, Philadelphia's "most gentrified neighborhood," where I lived my first year of graduate school but had to leave because my roommate and I could not afford the rent hike on the lease renewal (Philadelphia lacks robust renter protections).[154] I watched a tiny abandoned lot across from a new coffee shop become a "pocket park," now regularly hosting toddlers on its small playground. Larger vacant plots closer to my home in Point Breeze served as dog parks and thus de facto community gathering spaces, both for the area's African American residents and the newly arrived, wealthier white neighbors. In the midst of a massive housing boom, however, these sites of canine-mediated interaction would only last until fencing went up and construction crews started digging the foundations for new single-family housing, likely to feature the jarring and idiosyncratic "boxy bay" designs that have become iconic of Philadelphia's gentrification.[155]

Outside my immediate observations, institutional changes were also underway. As density increased and ride-hailing platforms Uber and Lyft began operations (deemed illegal but nonetheless tolerated for three years), traffic became noticeably worse on the city's mostly narrow, one-way streets. Delivery startups Caviar and Postmates launched operations, as did several other outlets—the locally owned goPuff, for example. Bike couriers and delivery vehicles were suddenly everywhere; riders with box-shaped thermal backpacks now cut in and out of lanes on busy streets, delivering meals to the city's growing ranks of tech "innovators," movers, and shakers. In 2011, the tech news organization Technical.ly held the first Philly Tech Week, an annual "open calendar of events celebrating technology and innovation in the Philadelphia region."[156] Then mayor Michael Nutter secured a large grant from Bloomberg Philanthropies' Mayors Challenge and issued an executive order to establish a Mayor's

Office of New Urban Mechanics. This new, privately funded government initiative would launch several web- and app-based products, including Textizen, a text messaging system for citizens to offer civic feedback, and Philadelphia2035: The Game, a web app for citizens to review and provide input on the city's twenty-plus-year planning document. All these initiatives would shutter within two years as the philanthropic funding dried up and a new mayor was elected.

Despite all this momentum toward Philadelphia's digital future, the city continues to hold the title of poorest big city in the United States.[157] Poverty hovers at above 25 percent, and the poor tend to be concentrated in large pockets that bear little semblance of their previous stature as working-class enclaves adjacent to industry. For example, located just a couple miles northeast of Center City, Kensington was once home to a sizeable chunk of Philadelphia's once-booming manufacturing sector; today, Kensington hosts the largest open-air drug market on the East Coast (sometimes called "the Walmart of heroin").[158] And while the median income has been steadily increasing around the country, in Philly it's been falling, undermining the city's "self-styled image as a revitalized metropolis on the rise."[159]

Perhaps nothing is more iconic of these contradictions than one particularly unsubtle development, Pennovation Works, the University of Pennsylvania's entrée into the startup incubator market. As a "strategic blend of offices, labs, and production space that pushes for the advancement of knowledge and economic development,"[160] Pennovation houses robotics firms, biotech startups, med-tech labs, and other bleeding-edge tech impresarios. Its layout follows from a "flexible framework" in the "spirit of ground-up innovation," its jagged design breaking free of the site's industrial past to encourage entrepreneurship and creativity (Figure 1). If all goes to plan, the campus will serve as the anchor to a brand-new "innovation district" for collaboration and financial partnerships between the university, other research institutions, corporations, and nonprofits.[161]

Pennovation occupies twenty-three acres along the Schuylkill River on the site of a former DuPont plant and research lab, where industrial scientists once invented new lacquers for quick-drying automotive paint. Before that, the plant was home to Harrison Brothers' Grays Ferry Chemical Company, another paint production operation employing upward

Figure 1. The main structure of the University of Pennsylvania's remote Pennovation campus and startup incubator. The site houses robotics firms, biotech startups, med-tech labs, and other academic–industrial partnerships; the building's architecture references the campus's industrial past and technological future as the anchor of a larger "innovation district." Photograph by the author.

of five hundred workers. A century's worth of heavy industry made the campus an anchor for nearby communities.

The area today sits the center of an infrastructural knot, an industrial zone both separated from and connected to the city and other regional centers by train tracks, river, and highway. The name of the small, predominantly African American residential cluster adjacent to Pennovation's campus, Forgotten Bottom, claims the area's isolation as a defining characteristic. Forgotten Bottom's footprint is small enough that it shares a census tract with the adjacent Grays Ferry neighborhood, which, like Point Breeze to the east, is fast becoming a quarter in the crosshairs of the city's mounting gentrification pressures. On Forgotten Bottom's west flank, the Pennovation site opens to an entrance onto the city's busiest artery, the Schuylkill Expressway leg of Interstate 76. To the southwest lays an expanse of trainyards that once serviced the city's massive, 1,400-acre oil refinery—the largest on the East Coast.

This unique combination of access and isolation makes the neighborhoods an attractive zone to developers, especially with Pennovation now in the mix. According to the 2017 American Community Survey, over 35 percent of Forgotten Bottom and Grays Ferry residents were at or below the poverty line; that rate was a mere 7.5 percent in the adjacent and far more integrated tract at the western edge of Graduate Hospital. With the flight of manufacturing from Philadelphia and other urban centers, disinvestment in neighborhoods like Grays Ferry and Forgotten Bottom was devastating; and that same abandonment now makes the neighborhoods ideal targets of financial and real estate speculation. In the summer of 2019, an explosion and fire at the oil refinery shook the ground, waking residents over two miles away and forcing Philadelphia Energy Solutions, the field's operator, to shutter. In fast-changing Philadelphia—in the midst of the largest construction and development boom the city's ever seen—it did not take long before housing developers started speculating about the bounty of riverfront land in the refinery's footprint, now unoccupied and adjacent to Pennovation, the anchor of the yet-to-be-built innovation district.[162]

I watched Pennovation's construction over several years. Even before the ribbon-cutting ceremony in 2016, it was obvious that the university would not be integrating the campus with the surrounding communities. The site's black metal fencing, as well as the visible omnipresence of security guards and surveillance cameras, made it clear to Pennovation's neighbors that getting onto the campus required university affiliation and proper identification.

On my way home on a bright and crisp Tuesday afternoon one November, I stopped in front of Pennovation at the intersection of Thirty-Fourth Street and Grays Ferry Avenue, observing its resident innovators at work. Below is the account from my notes:

> Over the black chain-link fence at Pennovation—beyond the security way-station and turnpike to the parking lot—two men stand under a canopy of netting adjacent to the main building, attending to what I would later learn is called a Minitaur—a four-legged, headless, leaping robot produced by Pennovation tenant Ghost Robotics for the military. The "ground drone," as it's called, can open doors and climb fences. Here, the robot is on its back. The roboticists are tinkering with something on what I'd be inclined to call its "belly." A couple twists of a tool and the Minitaur is back on its feet, insect-like and spry. I gawk through the fence at the alien machine for

> a few minutes before surveying the intersection. Across from one of the gas stations, an abandoned grain silo in an overgrown lot; a person's belongings in a shopping cart, placed behind a patch of brush for safe-keeping. A man, perhaps the owner of those personal effects, flies his sign in the middle of the five-lane intersection, asking drivers at the red-light for spare change. A woman—perhaps his partner?—is across the street and doing the same, her cardboard sign, facing me, reads ANYTHING HELPS. EVEN A SMILE .

As much as the property managers might wish it were the case, Pennovation does not exist in a vacuum. As our cities become "smarter," contrasts like these—between the bright techno-futures that Pennovation symbolizes and the material remains of an economy and social contract that we are leaving behind—will only become more common. We have to understand spaces like Pennovation not just in terms of their fortification and securitization, but through the support that they receive from multiple infrastructures. Yes, there are the adjacent networks of heavy industry, but there are also the social and technological infrastructures that sort and filter between the inside and outside of Pennovation's walls—all those networks keeping the people on the inside content to "innovate," and those on the outside out.

The next time I would visit Pennovation was when I was conducting research for chapter 2 as a bike courier for Caviar, shuttling food from a high-end, vegetarian, fast-casual takeout spot on the University of Pennsylvania's campus to the friendly engineer, programmer, or entrepreneur who met me at the outer gate. Pennovation's "innovators," it seems, like to take advantage of the city's now-competitive market for "digital transportation services."[163] With food-delivery startups like Caviar, Pennovation's engineers and entrepreneurs had no need to buy their lunches in Forgotten Bottom or Grays Ferry, which, until recently, qualified as food deserts. The area would come up again during my research for chapter 3, this time as a box on a crime-prediction map of Philadelphia in an image that the product team at HunchLab used to pitch their platform to police departments. Disinvestment and poverty, of course, motivate crime, but the crime map flattened this dynamic to a machine-readable topology of danger that obscures the neighborhoods' deep sense of community. By contrast, notably absent from the area were any of Philadelphia's recently launched LinkPHL Wi-Fi kiosks. An offshoot of New York City's Wi-Fi network that I study in chapter 1, LinkPHL's electronic screens now line the boulevards and avenues of Philadelphia's Center City business district. Forgotten Bottom and Grays Ferry were never part of the plans for the LinkPHL network. This, despite the fact that Qualcomm, a partner in the consortium that produces the Links, maintains a satellite office in Pennovation, and despite the clear benefits that free, high-speed internet access would bring to Forgotten Bottom and Grays Ferry—where the local Queen Memorial Library's computers are only available on weekdays until six in the evening. After all, with a network funded by advertising dollars, what private firm would be willing to shell out hundreds of thousands to construct and maintain Wi-Fi in an area where the residents have such meager buying power?

The smart city, as it actually exists today, is both Forgotten Bottom and Pennovation; it is the logistical networks and infrastructures that support their stark juxtaposition and separation. This is what I hope to communicate in the following pages—that it is only by a failure of our collective imagination and political will that we continue to see the uneven distribution of technology's benefits as a natural and inevitable outcome. What logics and techniques sustain this fantasy?

1

Design

Optimizing the Ad-Funded Smart City

There are structural asymmetries built into the very notion of a sensor society insofar as the forms of actionable information it generates are shaped and controlled by those who have access to the sensing and analytical infrastructure. . . . In general terms, a sensor is a device that measures or detects an event.

—MARK ANDREJEVIC and MARK BURDON, "Defining the Sensor Society"

We understand each Link's exact location and we understand time. And so, when you add those two things together, you get events.

—DAVE ETHERINGTON, Chief Strategy Officer, Intersection

A Technological Asset

In October 2014, New York City was at a telecommunications crossroads. The franchise agreement between the Department of Information Technology and Telecommunications (DoITT) and the thirty-nine companies operating the city's 11,412 payphones was set to expire. And while the end of a payphone franchise would not normally warrant much attention, in the years since the agreement began, the political economy, technology, and social norms of telephony had undergone a dramatic transformation. Cellphones had all but ended the payphone's reign as a staple of electronic communication in public spaces. In 1999, when the franchise agreement started, the number of public pay telephones in the United States was at an all-time high, with more than 2.1 million in operation; by March 2014, the year the franchise expired, that count had plummeted to a mere 150,000.[1] Over the same period, cellular subscriptions in the United States jumped from 30.6 for every hundred people to more than 110, meaning that a good chunk of the populace had more than one mobile device.[2] The Pew Research Center estimated in 2019 that upward of 96 percent of U.S. adults owned a cellphone, about four-fifths of which were "smart," internet-enabled, and location-aware.[3]

Despite the rise of the cellphone, payphones remained a persistent if neglected presence in New York City. Operators were actually still making decent money, even into the 2010s. It was just that most of that revenue was not coming from phone calls. The $1 million a year that operators brought in from coin-drop and dial-around payments was a pittance compared to the $50 million that they saw in advertising revenue.[4] Payphones, it turns out, were valuable marketing real estate. Millions of New Yorkers and visitors to the city pass by payphones every day. As long as the advertising was up, operators were happy, even if the phones themselves fell into disrepair. Aside from a few artists devising clever reuses for the neglected booths or enthusiast photo-blogs dedicated to payphones, the old structures quietly faded into the background of our urban environmental awareness—just another piece of noise in an already-loud visual ecology.[5]

This all changed in October 2012 when Hurricane Sandy thrust payphones back into New York's collective consciousness. After years of abandonment, New York City's payphones became a lifeline for the millions stuck in the city during the superstorm. The hurricane wrought mayhem on the coastal metropolis. Electricity was down across much of the city's five boroughs. The subways were flooded. Cellular connections were all but impossible to find. Payphones were the only option for New Yorkers to connect with friends and families, to coordinate healthcare or childcare. With so many booths out of order or in disrepair, however, local news outlets and websites had to send reporters out to find the few phones that remained in decent shape. "Of the 20-odd pay phones we visited, nearly all had some sort of malfunction," reported a local news site for Manhattan's Lower East Side. "Whether it was a lack of dial tone or gunk plugged in the coin slot, issues abounded."[6] Lines formed around the booths that did work.

While cellular service was eventually restored, many New Yorkers discovered a newfound appreciation for the aging payphones. And with the telecom franchise nearing its end, the City felt that it could not just remove the booths. That would be a slap in the face to the infrastructure that had just proved its disaster-proof worth. But at the same time, officials argued, it did not make sense to simply replace the aging structures with new phones. Payphones "are technologically the most valuable asset we have in the city," New York's Chief Information and Innovation Officer Rahul Merchant offered, "and it'd be wrong of us to replace one payphone with just another."[7]

What then does a city do with communications infrastructure that outlives its day-to-day relevance but cannot be removed? This chapter studies the afterlife of New York City's payphones as an iconic case of smart city retrofitting. Starting from the digital interface and the frictions it generates, the story I tell here refuses to fall back on corporate visions of a technological world to come. But I also take seriously the role that those visions play in a politics of technological anticipation—struggles to define a collective sense of excitement and inevitability around particular designs and features.[8] I track the payphones' transformation from humble "public communications structures" to LinkNYC, a gigabit-speed, public Wi-Fi network. LinkNYC's futuristic, nine-foot-tall digital kiosks—known as Links—come loaded with a suite of sensors, including Bluetooth beacons, microphones, and cameras, as well as USB inputs for phone charging, a touchscreen interface, and two fifty-five-inch LED screens dedicated to advertising. When completed, LinkNYC will not only be a first-of-its-kind citywide network; it will be the world's largest digital out-of-home advertising platform.[9] At the time of writing, some 2,000 of the projected 7,500 Links are now active across all five of New York's boroughs. Sister networks, operated by the same consortium of companies, have gone live in Glasgow, Newark, London, and Philadelphia.[10]

Although LinkNYC may today seem like an obvious choice to replace the payphones—after all, speedy Wi-Fi is a pleasure to use, and most of us have had an experience when a cellphone charge would be mighty convenient—back in 2012 and 2013, it was not at all certain what the "future of the payphone" would hold. Tracing LinkNYC's development allows us to glimpse how smart city interventions are shepherded from vision to reality. As science and technology scholars have shown, successfully designing, prototyping, developing, and marketing a new technology is tricky business. Technologists and designers must manage and negotiate between competing visions and conflicting expectations, ensuring that their technology seems inevitable while alternatives appear impractical or unreasonable.[11] All of this work certainly went into LinkNYC. The various actors involved, from the Mayor's Office and DoITT to the designers and technology companies competing to win the bid, had to garner support for their designs, demonstrate the value of certain features over others, balance the expectations of a diverse set of stakeholders, and manage undesirable uses after implementation.

To understand these efforts, we need to consider the context in which LinkNYC came to replace the payphones. LinkNYC launched in the midst of a marketing gold rush. As competition in online advertising heats up and users grow wary of platforms' surveillance and misinformation, marketers are turning to what has been pitched as advertising's final frontier—the physical spaces of our cities' streets.[12] According to the Outdoor Advertising Association of America, digital out-of-home advertising remains (at the time of writing) the only marketing sector to enjoy continued growth amid declining sales in traditional outlets.[13] Sales on networks like LinkNYC accounted for a combined $2.7 billion in U.S. ad spending in 2018, and the market only expects further increases.[14] Unlike traditional out-of-home marketing on, say, billboards or even posters on the backs of payphones, *digital* out-of-home advertising adds a feedback component by capturing data on the audience and their surroundings.[15] Not only do passersby see the ads—the ads see them.[16] This data capture is an incredibly exciting prospect to advertisers, first, because it allows marketers to *target* their messaging to particular audiences; and second, because it allows for the automation and optimization of ad *pricing*. LinkNYC may therefore have been celebrated as a first-of-its-kind citywide Wi-Fi network, but as a marketing platform it takes digital advertising to new heights. LinkNYC blankets the city in a "doubly communicative" mesh, its Wi-Fi and other services sending and collecting data in the same interactions.[17] Even passersby not connected to the Wi-Fi are subject to LinkNYC's data capture.[18] All this is incredibly value to an industry pegged to covert data collection.

Luckily for marketers, cities are ready and willing to trade their citizens' data for private infrastructural investment. Whether this data is useful for advertising is a different question. According to *TechCrunch,* the economics of "ad-funded smart city tech" only make sense if the data is viable for marketing purposes, and this requires *engagement.* The infrastructure must be "something with which citizens will engage—and engage a lot." You cannot expect that brands "will be willing to pay to throw their taglines" on just anything. "The infrastructure has to support effective advertising," which means "building out the network of sensors, connected devices and infrastructure needed to track city data."[19] If anything meets these criteria, it is LinkNYC. The network's free Wi-Fi and other offerings encourage users to "engage" the system for extended periods; its ubiquitous presence across New York's five boroughs means that it is hard to avoid; and its digital displays mean that ads can be tailored and rotated to target specific audiences. Ubiquity, addressability, and flexibility in a single package—the ideal conditions for an advertising-funded overhaul to the payphones.

This "intersection" of affordances is hardly lost on Intersection, the lead in the consortium of companies that operates LinkNYC. Intersection enjoys a dual identity as both marketing firm and smart city provider. It weds sensing to advertising, media sales to data brokerage. And the company's executives are hardly shy about the value of all the Link data. The kiosks are distributed in space and they record information in time, and as Intersection's Chief Strategy Officer Dave Etherington suggests in the epigraph, those details converge as *events.* To marketers, events mean profits. Events connect messages to contexts, products to consumer preferences and behaviors. In fact, Etherington envisions a whole new world of event-driven marketing opportunity—"to do more than serve advertising *near* specific places":

> Like, with restaurants or stadiums, we could tell you if Justin Bieber is playing, or—not just that a restaurant is there, but that there's a table for four available at a particular time.... With the data, we have an opportunity to really help our advertisers think about advertising as a brand-new

> medium, one that's filled with interesting data insights, one that enables you to serve ads when you see changes in context, changes in audience behavior.[20]

With this access to context and behavior, LinkNYC inaugurates one of digital advertising's most ambitious undertakings: to automate and optimize the buying, selling, and distribution of advertisements by translating the surveillant logics of online marketing into the physical spaces of our cities' streets.[21] "We want to build the type of advertising platform that's perhaps more analogous to advertising online or on mobile than what exists in physical spaces," Etherington explains.[22] Or, put more simply by Dan Doctoroff, CEO of Sidewalk Labs, a Google sister company and major investor in Intersection, to "replicat[e] the digital experience in physical space."[23] Such replication requires a calculative infrastructure (a way of seeing and analyzing urban data) and a responsive architecture (a material network that reacts to those calculations)—in short, a wholly logistical operation, and LinkNYC provides the ideal platform.

How then did New Yorkers get stuck with such a system—with an infrastructure that tracks their movements to sell ad space; a network whose furtive surveillance is routinely a target of activists' protests; and whose business model Nick Pinto of the *Village Voice* described as "sell[ing] citizens' privacy off the back of a truck to a for-profit company"?[24] Tracing LinkNYC's development reveals that advertising's Faustian bargain for the smart city—what media and technology critic Douglas Rushkoff called "a deal with the devil we really don't need"—both was and was not inevitable.[25] On the one hand, because the City tried to offload the costs of infrastructural overhaul onto its private partners, it yielded to industry interests and demands. Advertising's prominence on the structures was therefore all but preordained. On the other hand, because the administration sought credibility among key stakeholders and communities, the City offered a rare opportunity for citizen input. In 2012, the Mayor's Office and DoITT held a design competition to source ideas and concepts from the public, to encourage the "design and tech communities" to help the City "reinvent the payphone." Looking back to this design competition, it is striking how much more open and diverse visions for smart city infrastructure had been, even in just the recent past. Part of what the chapter does, then, is to excavate abandoned designs in an effort to "retrospect" prospective urban techno-futures—to reflect on alternative infrastructural imaginaries that might challenge our resignation to systems that promise public goods but whose business models rely on covert surveillance.[26]

LinkNYC is both an infrastructure and an interface, and all interfaces obscure at the same time that they reveal.[27] The risk is that a system like LinkNYC will obscure so much that we forget the obfuscation entirely and can no longer see the infrastructure. "We often have little understanding of how and where the mediation of urban systems takes place within the city itself," media scholar Shannon Mattern writes in her critique of smart city interfaces. "Nor do we know how *our* intelligence translates into urban 'sentience,' and what is gained or lost in the conversion."[28] Is it possible to "reverse engineer" this translation—to understand how obstruction gets systematically designed into our urban networks? To challenge the sanctioning of certain uses and users while others are policed as "abusive"? Tracing LinkNYC's design history allows us to see that the "ad-funded smart city" is not inevitable, nor is it "optimal." By juxtaposing the network against alternatives, we begin to see just how narrow, unimaginative, and non-inclusive the smart city is when motivated by the profit-driven logistics of ad circulation, and not public benefit.

From Blue Sky to Gray Reality

After Hurricane Sandy, payphones were all over the news. "Pay phones are suddenly important again because of Sandy," proclaimed Mark Memmott for National Public Radio, while the *Wall Street Journal* wrote that Sandy prompted "wired New Yorkers [to] get reconnected with pay phones."[29] With all this attention on the lowly communications structures, former New York City Mayor Michael Bloomberg saw an opportunity. It was 2012 and although the City was running trials to convert payphones to Wi-Fi hotspots, it would have been politically imprudent to remove the phones altogether. Like many of the administration's high-profile projects, Bloomberg wanted to leave his mark—not just by sprucing up the existing structures, but by wholly reimagining the payphone for the smartphone era. With a little creativity, Bloomberg's Chief Information and Innovation Officer Rahul Merchant predicted, the payphone infrastructure could again become "the city's most important public technological asset."[30]

In December 2012, just a month after Sandy, Bloomberg decided to make his excitement about the payphones known. He sent his Chief Digital Officer, Rachel Haot, to an NYC Tech Meetup at New York University where she would address the hundreds of "tech enthusiasts" in attendance. On stage, a video played behind Haot, with Bloomberg apparently speaking to her through a payphone mounted on an office wall adorned with a map of the city and a "Made in NYC" emblem. "From Wi-Fi in public spaces to the High Line," the mayor proclaimed in the prerecorded exchange, "our administration has continuously reinvented city infrastructure by matching innovative concepts with extraordinary designs. Now we're doing the same for the thousands of public pay telephones across the five boroughs, and we're challenging our dynamic and ever-growing

Figure 2. New York City Mayor Michael Bloomberg in a prerecorded video, addressing several hundred members of the NYC Tech Meetup group at an auditorium at New York University. In his message, the mayor issued a design challenge to the city's design and tech communities to "reinvent the payphone." December 2012. Photograph courtesy of Devindra Hardawar.

tech community to 're-own the phone' and provide their ideas on what the future of payphones could entail."[31]

The screen went blank and Haot elaborated on Bloomberg's message. The Department of Information Technology and Telecommunications was sponsoring a competition to source ideas and concepts for what the payphone of the future might look like. "It doesn't even have to look like a payphone," she explained. "We want to invite your prototypes and help build something that could potentially impact the future of payphone infrastructure in NYC."[32] The audience cheered. Many tweeted out their support the same night. One participant wrote, "I have never, and I mean NEVER been as amped about payphones as I am at this very moment. It's on. Come and #ReinventPayphones."[33]

News of the competition spread quickly through online networks. DoITT posted specifications to its website. Tweets flurried. In January 2013, a month after Haot announced the launch of the Reinvent Payphones Challenge, the administration held an information session for "students, urban planners, designers, technologists, architects, creators and legal and policy experts" interested in participating—anyone who wanted to "build physical and/or virtual prototypes imagining a new public utility through payphone infrastructure."[34] Haot again hosted the event, now accompanied on stage by a panel of DoITT staffers as well as experts in public space management from both the public and private sectors. The session, Haot explained, would be about communicating to the public the sense of opportunity that the design competition presented—not just for the winning entries, but for the entire city. "Our goal is, how can we set the global standard for what the future of urban connectivity can mean?"[35]

The project's ambitions were in the air, and they were infectious. But as the meeting got underway, it also became clear that Haot and the DoITT representatives needed to temper expectations. After declaring the competition's high stakes—setting a global standard is no small feat—Haot was quick to point out that winning entries would "not go on to change, formally, the city's payphone infrastructure. This is something that will follow the city's official procurement process and RFP [request for proposal] process."[36]

These and other statements began to set a more ambiguous tone. On the one hand, DoITT was interested in "blue-sky" ideas. "In the

#ReinventPayphones Design Challenge—from wifi to digital kiosks—the sky is the limit. Go beyond the phone," Haot tweeted when Bloomberg first announced the competition.[37] The City wanted to unleash and harness its citizens' talent and creativity. "Above all, this is about innovation and doing this in a collaborative way, because we've got the best and brightest here in New York City—at our design firms, at our academic institutions, at our tech companies, in the private sector and in New York City government."[38] But the blue sky did not quite square with the rather gray reality of city procurement regulations, RFPs, and zoning codes. On the other hand were the much more pragmatic constraints of municipal regulations—the tangle of often-conflicting codes that urban theorist Mariana Valverde argues work to secure political and institutional power at the expense of individual or collective agency.[39] "Obviously," Haot reminded the attendees, "we'll be constrained by the laws of what's allowed on the streets and what has to be on the streets, but aside from that, we are looking for ideas without constraint."[40]

The clarity of Bloomberg's excitement was getting muddled. The prospect of complex procurement requirements seemed to preemptively foreclose on citizens' genuine participation in the payphone's reinvention—let alone the implementation and rollout of whatever design eventually won the competition. Many in attendance seemed incredulous that finalists would not, in the end, have much say in the final design of the new structures. Audience members inquired about ownership arrangements, whether they could expect to retain rights to their design if they won. In general, attendees seemed far less interested in having their input "crowdsourced" than in directly authoring, in some way, the next generation of payphone. Participants continued asking if winning designs would receive credit or royalties, if designers would have to give up rights to the design or prototype. Haot's responses began to sound repetitive. "The idea is to be the spark for innovation, and again, this is completely distinct from any formal procurement processes for the City of New York."[41]

The disconnect between blue-sky creativity and the pragmatics of government procurement resonates with contradictions at the heart of what Marxist geographer David Harvey calls entrepreneurial urban governance.[42] Entrepreneurial governance realigns cities' commitments from their citizens to corporations. It favors initiatives to attract capital and wealth over the management and administration of welfare services.[43] For decades, entrepreneurial cities competed to attract investment in and from the FiRE sector (finance, insurance, and real estate). But following the 2007–8 financial crisis, the FiRE sector's stability came into question and cities began promoting a new sector—TAMI (technology, advertising, media, and information).[44] We saw this shifting entrepreneurial impulse, for example, when cities across North America scrambled to put together lucrative tax deals and incentive packages to attract Amazon's second headquarters.[45] As Harvey and others argue, such efforts nearly always result in a "race to the bottom," with city governments giving away benefits to already-wealthy transnational corporations, imposing unnecessary austerity measures at home and exacerbating uneven urban geographies of access and exclusion.

Entrepreneurial governance also affects everyday life in the city. As cities ramp up their efforts to attract capital investment, they create crises of authenticity that threaten the social and cultural foundations of the urban experience. According to sociologist Sharon Zukin, in addition to attracting capital investment, large-scale anchors of gentrification and real estate development—such as Baltimore's Inner Harbor or, more recently, New York City's Hudson Yards (a development that Bloomberg's administration spearheaded)—are designed to cater to the consumption preferences of tourists and resident elites, for instance, with high-end shopping districts and luxury hospitality services that remain out of reach to local communities. With such opulent consumer trappings, urban mega-developments become anathema to the kinds of racial, ethnic, and socioeconomic diversity that make cities exceptional to begin with. This "Manhattanization" of development, as Zukin calls it—"everything in a city that is *not* thought to be authentic"—exposes deep contradictions between high-end consumers' desire for authentic urban experiences and the homogenizing effects of real estate investment and development.[46]

Nowhere are these contradictions more warmly embraced than in the popular economic governance philosophies evangelized first by Charles Landry in London, and then by Richard Florida and his followers across U.S. cities. Florida's now-famous "creative class" thesis insists that cities can catalyze urban regeneration, attract investment, and boost job growth by "attracting and retaining" cultural capital—artists and knowledge workers.[47] Florida's ideas proved particularly attractive to small- and medium-sized city governments across North America and Europe as they struggled with the harsh realities of the post-Fordist city: empty storefronts and factories, a hollowed-out middle class, sputtering investment and development.[48] To these seemingly insurmountable challenges, Florida offered a playbook: first, city and state governments legislate tax benefits and incentive packages to attract FiRE- and TAMI-sector firms; these investments then bring knowledge and creative workers, whose preferences for edgy consumption (hip restaurants, galleries, music venues, and so on) pump cash into economically marginalized areas, typically artist enclaves at the fringe zones of gentrification.[49] All the city needs to do, apparently, is to support the cultural ecosystems in these zones, either directly or by coaxing corporate sponsorships. As Zukin's work shows, however, this inflow of cash has a knock-on homogenization effect that makes authenticity a perennial problem for cities following Florida's playbook. While members of the "creative class" crave genuine experiences, the financial speculation that their entrance attracts quashes the conditions necessary for "authenticity," even in the most superficial sense.

In the face of these contradictions, city governments and private developers try to bolster their credibility by *simulating* authenticity. Inviting public participation in urban planning and design processes is one way of doing just that, whether in the form of citizen councils or design competitions like the competition to "reinvent the payphone." Writing in 1990, urban design scholars Tridib Banerjee and Anastasia Loukaitou-Sideris celebrated the public participation inherent to design competitions. "There is something very open, civic-minded, and public spirited about the design competition," they observed. "It catches the fancy of lay citizens, draws attention of the news media, engages interests of potential donors and philanthropists, stimulates young designers to devote their creative talents to developing innovative ideas, and so on."[50] Of course, all these characteristics sound like net benefits, and they certainly address some of the credibility issues stirred up by "Manhattanized" development. But as sociologist Gordon C. C. Douglas recently found, "participatory" design and planning can easily become enmeshed in the same struggles over larger economic forces of gentrification that they are meant to ameliorate. The rhetoric of participation, in short, "authenticates" inequitable development.[51]

Throughout his three terms as New York City's mayor, Bloomberg clearly grasped the connections between corporate tax incentives, tech-sector development, legitimacy, and participatory design, even if he failed to acknowledge the repercussions. After the financial crisis of 2007–8, Bloomberg began directly targeting his investment incentives at the tech sector. The ambitions are evident, for example, in *A Road Map for the Digital City,* a strategic planning document that Bloomberg commissioned in 2011 and which argues that "achieving New York City's digital future" must involve "crowdsourcing and contests" to "cultivate public participation."[52] The design competition to reinvent the payphone was likely the largest of these crowdsourcing events, but it wasn't the first. The administration had already undertaken similar initiatives to bolster support for large-scale real estate developments, most notably the High Line, a raised park on a stretch of defunct rail on Manhattan's West Side flanked by glitzy new skyscrapers. In the High Line example, we see how Bloomberg's particular variety of participatory design obscures the harsh political–economic reality blooming around the park. At the same time that Bloomberg held a competition for architects to design the High Line, the administration was undertaking cunning and sometimes draconian measures to spark real estate speculation in the areas adjacent to the proposed site.[53]

Participation thus lends the cutthroat world of real estate speculation and investment an air of authenticity, local ingenuity, and inclusiveness. Design competitions perform a certain humility among city agencies, with experts appearing to yield their authority to lay ideas. The *Digital Roadmap,* for example, suggests that "public challenges and contests cultivate creativity, involve New Yorkers in City projects, and introduce fresh thinking into government."[54] Haot certainly made this sense of humility front and center for the Reinvent Payphones Challenge. "What will come of this will be a lot of designs and ideas that people sitting at desks in city offices wouldn't have thought about."[55] But as we will see, it is never clear that the City ever took the designs, prototypes, and ideas submitted to the contest very seriously. To use planning theorist Sherry Arnstein's terminology, Bloomberg's brand of institutionalized participation was entirely tokenistic. It "allow[ed] citizens to advise or plan ad infinitum but retain[ed] for powerholders the right to judge the legitimacy or feasibility of the advice"—a superficial nod to the public side of the public–private partnership that performs humility while preserving authority.[56]

Moreover, involving the public in municipal projects has publicity benefits. As Banerjee and Loukaitou-Sideris made clear decades ago, design competitions do more than merely garner community input. They also attract attention through their spectacle, both among professional communities of designers and architects and the public at large.[57] Politicians and developers are not likely to overlook these benefits. According to urban designer Ute Lehrer, "Using [design competitions] as a vehicle to draw public attention to a project—and turning the process into a spectacle—appears to become sometimes as important as the outcome."[58] Through the tokenistic spectacle of participation, the design competition becomes an opportunity for powerful actors to appropriate the public's creative labor. This was especially true for New York City under Bloomberg, a mayor who staked his legacy on "remak[ing] the city's landscape for the 21st century" and was known for his cozy relationship with developers.[59]

Variations on a Well-Established Theme

The Reinvent Payphone Challenge thus fit squarely within Bloomberg's modus operandi. But it was also true that to many participants and observers, the competition *felt* grassroots—a rare opportunity for "techies, urban designers, and policy wonks" to have a personal impact on their city's future. The competition was billed as the first of its kind in the world—a "bottom-up" approach to what are otherwise almost always "top-down" procurements, acquisitions, and specifications, imposed from on-high by city governments or their corporate partners.[60] The Ash Center for Democratic Governance and Innovation at Harvard's Kennedy School recognized Bloomberg for the Reinvent Payphones Challenge with its "Bright Idea in Government" award, lauding the administration for inviting "hundreds of urban designers, planners, technologists and policy experts to create physical and virtual prototypes that reimagine the future of the City's aging public pay telephones."[61]

With all the excitement around the design competition—the write-ups on tech blogs, the awards, the back-and-forth between Haot, DoITT, and the design and tech communities—it certainly seemed like the City was planning to do something with the crowdsourced designs. Publicly available documents contradict this straightforward narrative, however. Well before the design competition was ever announced, DoITT was already honing specifications for the payphones' replacement. Five months before Bloomberg issued his challenge to "re-own the phone" at the NYC Tech Meetup, DoITT officials issued a request for information (RFi) to the current payphone franchise operators, soliciting advice about the structures that would eventually replace the payphones. Reading the RFi today, it is clear that city administrators—all those "people sitting at desks in city offices," as Haot described them—already had a well-formed idea of the kind of structure that could replace the payphone, down to specific design features:

> Free wi-fi service in public spaces has been widely discussed in recent years as a desirable amenity for pedestrians and other users of public spaces. . . . What would be the effect on City franchise revenue of incorporating such a network of antennas offering free service? . . . Could touch-screen technology make it possible, for example, for a digital advertising panel to convert to a neighborhood map or a subway map at a pedestrian's touch? Would it be desirable and practical to offer powered mobile device plug-in? . . . Can a two-sided, single-panel design with a relatively slim profile replace the current three-sided, curb line payphone enclosures as a primary design to include useful communications sidewalk amenities, possibly including some payphones?[62]

Telecom operators' responses to this inquiry (also publicly available) reveal that replacing the old structures with digital signage was an enticing prospect. New York City is the largest outdoor media market in the United States, with ad spending nearly twice that of Los Angeles, the second largest.[63] And to the city's payphone and "street furniture" operators (including Cemusa, JCDecaux, CBS Outdoor, and others), going digital would yield a major expansion of sales opportunities. Unlike paper posters, digital displays can be changed over or "playlisted" between ads at little to no cost; the operator can therefore cater to multiple clients on the same screen on a short rotation, bringing in a greater share of revenue per hour.[64] For example, in its response to the RFi, Titan Outdoor, the largest payphone operator in the city after having purchased Verizon's payphone stock in 2009,[65] suggested that the addition of digital signage would more than compensate for the decreased ad space on a two- rather than three-paneled structure. While "the loss of the third panel would significantly diminish the revenue potential of the New Franchise" because it "removes the lucrative 'street' panel, which is very attractive to advertisers," this "loss of potential revenue could be solved by allowing the installation of digital panels that would increase revenue due to the yield associated with digital advertising."[66]

DoITT heeded the operators' advice. Reading across the 2012 RFi and the 2014 request for proposals (RFP) reveals significant overlap in language and specifications. Where the 2012 RFi asks whether "a widespread network of wi-fi antennas incorporated into sidewalk panels [would] be desirable and feasible," the 2014 RFP states that "each Franchise Structure with advertising on it must provide, in addition to telephone service, completely free wi-fi (wireless fidelity) service." Where the RFi asks "what types of potential communications sidewalk services should be provided to pedestrian-users free of charge and what types should be paid for by such users," the RFP states that the franchise winner "will not be permitted to charge a fee for wi-fi service but may charge a fee for phone services with the exception of 911 and 311 phone calls" and will have "the option to install cell phone charging stations and permission to charge a fee for cell phone charging station services."[67]

These similarities disabuse the notion that the City ever took the designs submitted to the competition very seriously. Either DoITT simply dismissed the public input altogether or its preferred specifications (publicly available in the RFi) influenced the prototypes submitted to the Reinvent Payphones Challenge. We are unlikely to ever learn the full story. When I contacted DoITT, the agency's legal team determined that all but one of my questions involved confidential information; the agency's commissioner told me only that staff at the Department of City Planning "did examine the winning designs and may have been influenced by them."[68] What, then, did the winning designs look like?

If the RFi and RFP were in lockstep, so, too, were the competition judges. Nearly every entry selected as a finalist converged on the same or similar set of design features. One observer even described the finalists as "variations on a well-established theme."[69]

Entries were due in February 2013. A panel of judges selected eleven of the 120 or so submissions as semifinalists across five categories: connectivity, creativity, visual design, functionality, and community impact. The now-defunct New York City startup Quirky hosted a demo day, with semifinalists presenting their prototypes before an audience of DoITT employees and members of the press. Six were chosen as finalists, one for the first four categories and a tie in community impact. Images and descriptions of the entries were then uploaded to Facebook, where observers could vote for a "popular choice" winner.

Figure 3. A digital rendering of NYFi, the boutique New York City design firm Sage and Coombe's entry to the Reinvent Payphones Challenge and winner in the connectivity category. The design resembles specifications laid out in the New York City Department of Information Technology and Telecommunications' "Request for Information [Regarding the Future of Public Pay Telephones on New York City Sidewalks and Potential Alternative or Additional Forms of Telecommunications Facilities on New York City Sidewalks." Image by Sage and Coombe and presented by the NYC Mayor's Office (https://www.flickr.com/photos/nycmayorsoffice/8535340696/in/album-72157633214587930/).

The popular vote came in for an entry called NYFi, the finalist in the connectivity category (Figure 3). Submitted by boutique design and architecture firm Sage and Coombe, NYFi abandoned the traditional three-sided payphone structure for a sleek two-paneled monolith with large digital screens, one dedicated to app-based services (such as Yelp), the other to advertising. The concept drew explicitly on the minimalist aesthetic of the Apple iPhone—as the designers put it, "to help reduce the amount of clutter on the street and reinforce the Apple-inspired minimalist feel."[70] And in this, NYFi was not alone. Aside from Windchimes—the only finalist to augment rather than replace the payphones, and the only entry submitted by students unaffiliated with a professional design or architectural studio—each of the winning designs drew on the app-based mobile operating systems of the Android and iPhone, now stretched to the proportions of a phone booth. Like NYFi, Frog Design's Beacon (Figure 4) featured a slender, twisting two-panel display, again with large digital signage dedicated to advertising and an app-based interface. NYC

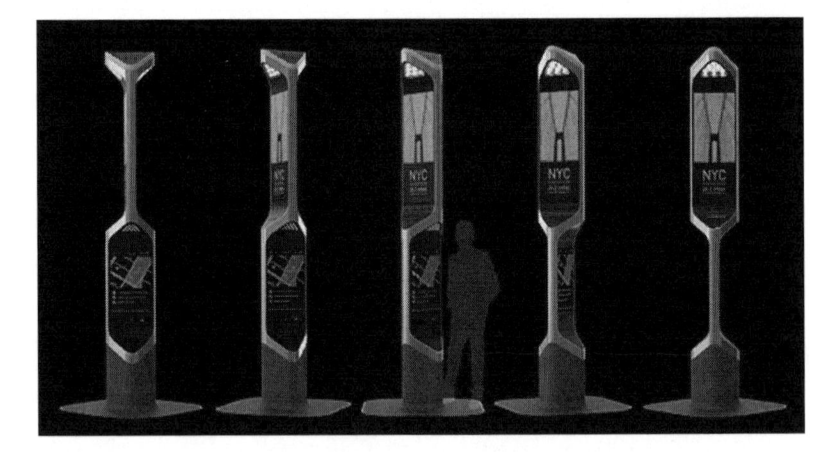

Figure 4. Frog Design's entry, Beacon, was the finalist in the Reinvent Payphones Challenge's visual design category. After the competition, the design captured the attention of LQD Wifi, "a technology startup with the mission to make Smart Cities real through ubiquitous wifi." LQD was later acquired by Verizon as part of the latter's investment in smart city infrastructure and later renamed Verizon Digital Kiosk. Image by Frog Design and presented by the NYC Mayor's Office (https://flickr.com/photos/nycmayorsoffice/8534232365/in/album-72157633214587930/).

Loop by architects FXFOWLE featured a wavy booth-like structure, with the internal "touchscreen phone" connecting to an external bench. Smart Sidewalks, submitted by a team of professional and academic architects and designers, included a touchscreen extending from LED strips embedded in the sidewalk. And NYC I/O, entered jointly by payphone operator Titan Outdoor and tech designers Control Group, envisioned a transparent, semicircular glass booth to house a touchscreen panel of apps for city services (Figure 5).

While each winning entry may have had its own "hook," they all combined a near identical roster of functionality: touchscreen computing, interactive maps, location-based services, USB ports for cellphone charging, traditional telephone calls, emergency services, and a Wi-Fi hub. Although it was never asked of them, all but Smart Sidewalks and Windchimes reserved space explicitly for advertising.

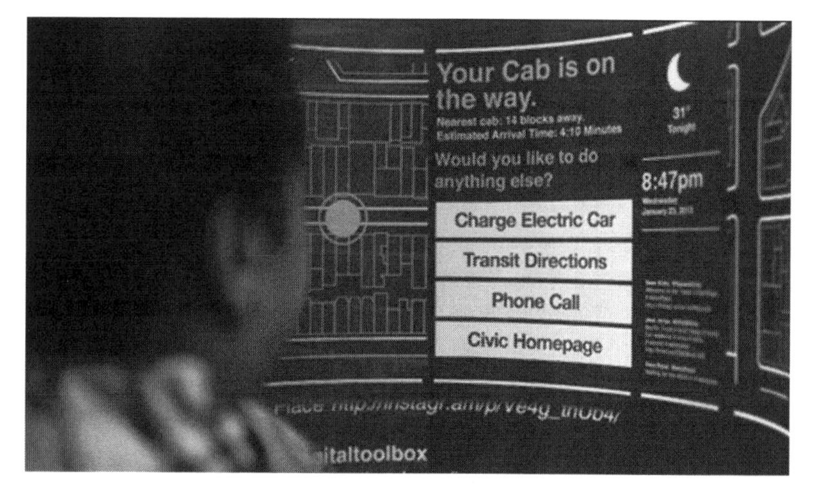

Figure 5. Payphone operator and media firm Titan Outdoor and design startup Control Group submitted NYC I/O jointly and tied as finalists in the community impact category. The concept mimics features of traditional phone booths while replacing its surfaces with a curved digital touchscreen to evoke futurity. Titan and Control Group would later merge to form Intersection, the lead in the CityBridge consortium that in 2014 won the franchise bid to replace the city's payphones with LinkNYC. Image by Titan and Control Group and presented by the NYC Mayor's Office (https://www.flickr.com/photos/nyc mayorsoffice/8534232285/in/album-72157633214587930/).

Anthropologist Lucy Suchman and her colleagues Randall Trigg and Jeanette Blomberg argue that design prototypes like those submitted to the Reinvent Payphones Challenge become "successful"—that is, they motivate product development or, later, achieve widespread user adoption—when they are both "intelligibly familiar to the actors involved, and recognizably new."[71] The competition finalists drew on the smartphone to achieve this intelligibility, invoking the iPhone or Android's familiar look and feel while reconfiguring its relationship to the human scale in recognizably new ways. Geographer Samuel Kinsley's work on speculative designs for ubiquitous computing corroborates these observations. Regardless of whether the design is a physical mock-up or a diegetic representation, prototypes come to function as "anticipatory objects" by invoking the "proximate future" through a "tension between imagination and practice," the new and the familiar.[72]

The Reinvent Payphones finalists work this way. As anticipatory objects, the designs invite their audiences to imagine using the reinvented payphones "spontaneously and effortlessly as they go about their everyday lives."[73] Emplaced in common urban settings, they induce "anticipated sensations of technological experience based on pre-existing bodily knowledge."[74] Moreover, if we follow Kinsley's argument to its conclusion, we see how the designs become embroiled in what he calls the politics of anticipation. Prototypes articulate a "semiotics" of anticipation. They give the *forms* (features, formats, specs, aesthetics) and *forms of action* (embodied interactions, haptic affects, tactile imaginations) that most effectively "evoke and produce futurity."[75] Like all symbolic systems, these semiotics refract uneven distributions of cultural authority and influence.[76] But the politics of anticipation is less about directly wielding that authority and influence than it is about managing collective *desire,* stimulating expectation as a source of anticipatory pleasure to shore up authority and influence. When certain features become mutually desirable to different groups—when the pleasure of anticipation becomes collective—"communities of anticipation" coalesce around the prototype, accruing momentum and cuing new participants into its technical discourse.[77]

The politics of anticipation, then, is a contest of attraction and momentum. And in smart city interventions like the Reinvent Payphones Challenge, the politics of anticipation collides with urban politics, with anticipatory communities of asymmetrically agentic actors coalescing around certain forms and forms of action but not others. Keep in mind the competition's stakes as a spectacle of authenticity. It would be hard to argue that the players involved—administration officials, DoITT staffers, tech startups, members of the press, professional designers and architects, and so on—were disinterested or evenly matched. All were situated figures with diverse motives vying to define the semiotics of urban futurity—to consolidate communities of anticipation around particular forms and forms of action. For one, the finalists—mostly professional designers, architects, and technologists—garnered praise for their work, and some used the competition as a platform for debuting marketable products. Frog Design's Beacon entry, for example, caught the attention of Verizon's smart city division and has since been rebranded as the Verizon Digital Kiosk.[78]

More fundamentally though, the Reinvent Payphones Challenge provided an opportunity for Bloomberg's administration to cultivate a sense of desire and inevitability around the features that DoITT wanted installed on the new communications structures. In the smart city, technology design is an urban question, and the judges' apparent consensus on the reinvented payphone's specs indicates the emergence of a lingua franca in the semiotics of the future city. This is not to say that City administrators or designers had some conspiratorial motive to converge on the smartphone and its intelligible familiarity as inspiration for the new structures, or that the designers intentionally appealed to the specs laid out in the RFi. But converge they did, and it was these "winning designs," not others, that "may have" influenced the Department of City Planning when it drew up the RFP for the franchise bid.

Ultimately, the gray reality of procurement and implementation not only ensured that already-vetted features would be included in the payphone's redesign. They also foreclosed on some of the more radical ideas submitted to the competition—designs that, as we will see later, could have done far more than LinkNYC to expand New Yorkers' access to urban spaces and infrastructures.

Prototyping at Scale

Not much happened in the months following the Reinvent Payphones Challenge. It was a full year before DoITT released the request for proposals, the official bid to "Install, Operate and Maintain Public Communications Structures in the Boroughs of the Bronx, Brooklyn, Manhattan, Queens and Staten Island." In the interim, Bloomberg's third and final term as mayor came to an end, and Bill de Blasio, who ran on promises to reduce the city's widening gap of economic inequality, was elected and sworn into office. De Blasio inherited the institutional legacy of the Reinvent Payphones Challenge, but he also sought to make the effort his own. The new mayor shifted the rhetoric to match his campaign commitments.[79] The RFP reflects this shift by prohibiting the new franchise holder from charging a fee for Wi-Fi service. If the operators wanted access to the ad space, they would have to finance connectivity. Rather than merely connecting "innovative concepts with extraordinary designs" as Bloomberg had promised, de Blasio insisted that the new infrastructure would become a great equalizer—a bridge over digital and socioeconomic divides.[80]

De Blasio's administration also tried to ensure that the City got its fair shake from the new deal. Under the 1999 franchise, 36 percent of payphone ad revenue went to the New York City government. De Blasio's 2014 RFP demanded a fifty–fifty split.[81] While this might seem like a steep price hike to potential operators, two factors ensured that the agreement would be attractive. First, unlike the 1999 agreement, which opened up the local telecommunications market to competition, the 2014 franchise would be awarded to a single firm. At stake then was nothing less than a monopoly hold on a citywide and government-commissioned network in New York's massive and lucrative out-of-home advertising market. Second, marketers believed that digital advertising was considerably more profitable than static billboards, and the RFP allowed for digital signage.

On the whole, the terms seemed like a win–win–win: the City would boost its revenue without having to raise taxes; de Blasio would score political points for providing New Yorkers free internet access; and the franchise operator would have exclusive access to the new structures' digital signage.

In November 2014, de Blasio announced LinkNYC as the winner of the bid, awarding the contract to a consortium of companies called CityBridge. At the time, CityBridge was led by payphone operator Titan and design firm Control Group (the team that submitted NYC I/O to the design competition), which were joined in the partnership by telecom giant Qualcomm as well as Civiq Smartscapes, a Comark company specializing in smart city infrastructures. CityBridge would be responsible for the costs of installing at least 7,500 Link kiosks and laying the hundreds of miles of fiber-optic cable necessary to connect the kiosks to the internet, a large chunk of which would run through Manhattan's complex underground conduits. CityBridge would also pay for operations, maintenance, and upgrades to LinkNYC's software and hardware, with a guaranteed Wi-Fi connection speed of at least one gigabit per second (almost twice as fast as the national average for residential internet).[82] In total, CityBridge estimated that it would spend $300 million for the first year of operations, none of which would come out of the City's budget.[83] It also promised to deliver at least $500 million in advertising revenue over the first twelve years—more than double what the government would have received under the previous franchise.[84]

Some observers lauded the City for its ability to dictate favorable terms in the new franchise, especially after years of failure in getting broadband into the outer boroughs.[85] But the deal's exclusivity also raised eyebrows. While Titan Outdoor may have been the largest payphone provider under the previous franchise, it still faced competition. Other operators called foul on the new agreement's monopoly-like arrangement. In December 2014, less than a month after the de Blasio administration announced that CityBridge was selected for the contract, the Franchise and Concession Review Committee received two letters of opposition, one from Cemusa, New York City's street furniture franchisee; and the other from lawyers representing Telebeam, one of the thirty-eight other payphone operators under the 1999 agreement.[86] The Telebeam letter made a number of allegations to discredit the CityBridge bid, including charges of favoritism based on personnel crossovers between Control Group, DoITT, and the city's Technology Development Corporation when the contract was under review.[87] The letter also cited evidence of Titan's poor standing with the city, as the payphone operator had failed to pay the Metropolitan Transit Authority $20 million in ad revenue from a separate agreement.[88] Most significantly, the letter claimed, the contract violated the 1996 Telecommunications Act's prohibitions on telecom monopolies.[89]

Members of City Council also took issue with the agreement, but for a different reason. Representatives from poorer, outer-borough districts that lacked commercial broadband access critiqued LinkNYC's geography of access. As required by the RFP, CityBridge developed two design options for LinkNYC, one for high-traffic areas with large digital ad signage and another for residential and historic-designated areas with little to no space for advertising. The council members were concerned that too few Links would go into areas where advertising was deemed unprofitable, or, if a kiosk was installed, that internet speeds would be compromised. The *New York Daily News* reported in November 2014 that Wi-Fi speeds in poorer neighborhoods in Brooklyn and the Bronx could be up to ten times slower than the ad-laden kiosks in Manhattan.[90] Reporters at *The Verge* calculated that even if each kiosk had maximum Wi-Fi speed, the spacing of the Links in the outer boroughs would be so sparse "that residents might as well just go to a coffee shop or library for free internet."[91] These potential disparities were actually consistent with the RFP. This led council members from the Bronx, Brooklyn, and Staten Island to file Resolution 0838-2015 "condemning the inequitable contract between the City of New York and LinkNYC" for providing "better wi-fi access to areas throughout the City where advertising is deemed more profitable."[92]

Despite these formal protestations—and despite a lawsuit eventually brought by Telebeam against the city[93]—the scale, scope, and exclusivity of LinkNYC attracted attention from big-name investors. The opportunity was too good to pass up. In June 2015, before ground had even broken on the first kiosks, CityBridge received a major financial boost. The capital came from Sidewalk Labs, a subsidiary of Google's holding company Alphabet that was launched only two weeks earlier. While Sidewalk would later become known for its controversial Quayside "smart district" on twelve acres of lakefront property in Toronto, the company's very first act was to secure a leading share in Titan and Control Group and merge the two to form a new company, Intersection, which would take the reins as the lead partner in CityBridge and LinkNYC's administrator.[94]

With Google's failed Wi-Fi network in San Francisco and its stalled Fiber venture in Kansas City, Sidewalk Labs would be the latest in a string of attempts by Google affiliates to invest in connective infrastructure as a way to secure digital advertising rights. The new firm brought Silicon Valley–style rhetoric with its investment, but Alphabet CEO Larry Page was also canny enough to know that conquering New York City's sidewalks required real estate and development expertise. Page tapped a player familiar to New Yorkers to serve as Sidewalk's founding CEO—Dan Doctoroff, the former Bloomberg Media executive turned deputy mayor of economic development and rebuilding, who oversaw nearly three hundred development initiatives for the Bloomberg administration. But Doctoroff's crowning achievement was Hudson Yards, the largest private real estate deal in U.S. history: a tangle of skyscrapers constructed over the train yards on Manhattan's West Side.[95] And in a remarkable show of corporate camaraderie, the newly formed Intersection's headquarters would be housed in the same building as Sidewalk Labs, adjacent to the Bloomberg's prized High Line, in Doctoroff's shimmering new Hudson Yards skyscrapers.

A Computational Turn

With the investments and leadership now secured and Intersection deputized to administer the network, the City held a ribbon-cutting ceremony in February 2016 to celebrate LinkNYC's launch. Within weeks, new Links were popping up along Third and Eighth Avenues in Manhattan; within months they would fan out through the island and into the outer boroughs. Per the franchise agreement, the entire system of nearly 7,500 Links would be up and running within just a few years, creating the world's largest public Wi-Fi network and digital advertising platform.[96] And as promised, the first Links tested at "blazing fast" speeds.[97]

Almost as soon the network was online though, LinkNYC stoked new anxieties. Many observers were wary of the public network's management by private entity. A month after the ribbon-cutting, the New York Civil Liberties Union (NYCLU) sent a letter of concern to de Blasio's counsel.[98] The letter praised the city's efforts to bring free, high-speed Wi-Fi to public spaces, but it also outlined a number of disturbing privacy implications—the network's ability to gather, retain, and share users' private information. This was not lost on Intersection. LinkNYC actually offered two Wi-Fi networks: one with significantly greater encryption and security, the other open and thus susceptible to malware and information theft. To access the secure network, users needed to register an email address and opt in to an agreement giving Intersection access to their web traffic. This meant that users were forced to choose between exposing their personal information to the company or opening themselves to the potential for online fraud and hacking. The NYCLU also raised concerns about law enforcement access to data. LinkNYC's privacy policy provided very little clarification as to whether and how police could requisition data or if municipal, state, or federal agencies would have access to the Links' video feeds. The sheer volume of data that LinkNYC generates, the NYCLU warned, "will create a massive database of information that will present attractive opportunities for hackers and for law enforcement surveillance, and will carry an undue risk of abuse, misuse and unauthorized access."

Intersection eventually updated its privacy policy to address some of these concerns, specifying, for example, when and how the company would hand over its video feeds to law enforcement. But the updated policy still failed to address the more furtive modes of surveillance that the network's design made possible. As a number of observers have noted, LinkNYC has the capabilities to capture data on any mobile device, *even those not connected to the network.*[99] The new policy prohibited the collection or sale of "personally identifiable" information (name, email address, phone number, and so on), but left "technical information" open to collection, retention, analysis, and third-party access. This would allow network administrators to record the presence of Wi-Fi-enabled devices—smartphones, laptops, tablets—within range of any Link simply by logging the device's MAC address, the unique numeric identifiers broadcast when attempting to connect to a Wi-Fi network. As the privacy policy states, "We know where we provide Wi-Fi services, so when you use the Services we can determine your general location." When the network is complete, Intersection and its partners will be able to track, in real time, the "general location" of nearly every Wi-Fi-enabled device in the city.[100]

What will all that data be used for? Intersection has refused to disclose its plans for the data. After *The Intercept* reported that an undergraduate student discovered tracking code buried deep in LinkNYC's programming, Intersection declined to comment on the code's purpose or whether and how the company uses location to boost ad sales.[101]

We do know, however, that the advertising industry places a premium on location, first, because it facilitates personalized ad *targeting*, and second, because it allows advertisers to collect *attribution metrics* that boost advertisements' value by dynamically adjusting prices where and when it is busier.[102] Both targeting and attribution are defining features of "programmatic" advertising, "the use of data and networked computing to automate and optimize aspects of media buying and selling."[103] This has long been a pipe dream of marketers and media buyers—a fantasy of perfect alignment between advertisements and their effects, media content and media audiences. In the era of network television broadcasts and mass communication, realizing that the programmatic dream required heavy equipment, infrastructural investment, and resource expenditure. Companies like Nielsen staked their business on measuring audience size and composition, a process that involved intensive surveillance (generally with audience consent) to create feedback mechanisms. With the rise of the dot-com internet, however, programmatic became lighter, its data collection more furtive. The medium (the internet) delivered both the message and the feedback mechanism; the network served up both content and audience surveillance.[104] Google's purchase of early online ad brokers, such as Applied Semantics, DoubleClick, and Adscape, made the search engine's ad-driven business model an industry leader in this regard.[105] As executives at Sidewalk Labs and Intersection make clear, LinkNYC is meant to replicate Google's business model in physical space, and its ability to collect location data is the cornerstone of the entire operation. Location is what makes Intersection's targeting and dynamic pricing possible—what the company calls "true real-time bidding" and "granular device-level attribution." When executives promise the "digital experience in physical space," then, they are not speaking to Wi-Fi users; they are talking to *media buyers*, Intersection's real clients, who can now "buy and measure media in the same way as they do web, mobile, and other digital media, and enjoy the benefits of massive reach, high-impact, always-viewable messaging that reaches consumers in the physical world. All with no bots, no ad blockers, and no unsafe content."[106]

With a New York City–sized market, Intersection is now an industry leader. But the company is hardly alone in its ambitions. The business model reflects a much deeper and far-reaching trend that critical communications scholar Jeremy Packer identifies as a "computational turn" in media–industrial logics. According to Packer, the only thing that now matters to advertisers are "directly measurable results"; not aesthetics, style, or ideology, but calculation, modulation, correlation, velocity—in other words, the *logistics* of ad circulation.[107] And it is no coincidence that Packer illustrates this point with Google's advertising business. Google is not concerned with messages' content or style or substance: "The *effect* is the content. The only thing that matters are effects—did someone initiate financial data flows, spend time, consume, click, or conform?"[108] Creative departments may still work behind the scenes to craft images and sounds capable of inducing consumer attachments to brands, but the issues that most immediately move marketers are location, timing, targeting, profiling—which ads to show to whom, when, for how long, and where?

Of course, these logistical concerns are not new to advertising either.[109] The gold rush to translate Google's effects-driven ad business to out-of-home marketing has, as John Durham Peters might put it, pushed advertising's logistics back to center stage.[110] Behind the scenes, the advertising industry's "math men" have long been focused on questions of circulation and audience profiling.[111] Only now it is a question of expanding those logics outside prime time, outside the web banner, out of the home—to provide evidence to media buyers not only that their billboard is in the optimal location to reach "eyeballs," but how long those eyeballs were on the message, whom those eyeballs belong to, the probability that the eyeballs' owners will make a purchase, and so on. All of that information increases the value of the marketing event. The larger the audience, the more advertisers can charge to display the message; the higher the probability of purchase, the higher the price.

Out-of-home advertisers are rushing to figure out ways to capitalize on all that surplus, to concoct audiences from city dwellers' day-to-day comings and goings. The trouble was finding a network large enough, flexible enough, and capable of collecting enough data to accommodate programmatic advertising's calculative infrastructure and responsive informatic architectures. Before LinkNYC, such a network did not exist (at least under the command of a single administrator like Intersection). And unless Intersection unexpectedly decides to embrace a radical model of corporate transparency, the full extent of LinkNYC's audience surveillance and calculations will remain a secret.

We can, however, draw some inferences about LinkNYC's operations. Computer scientists, for instance, have long been interested in calculating optimal pricing strategies for advertising networks, and these models reveal how a system like LinkNYC likely works. Such was the research problem that Jörg Müller, Alex Schlottmann, and Antonio Krüger of the Institute for Geoinformatics at the University of Münster set out for themselves in a model called "Self-Optimizing Digital Signage Advertising." This system, remarkably similar to LinkNYC as a network of digital ad displays, trains an algorithm on "the number of users looking at the display, if possible their identity, and the current time and location." It then uses that information to "execute a Bayesian network" that estimates "the probability that the users will act"—that is, that users will make a purchase based on their exposure to the advertisement. With this estimated probability, the network then "compute[s] the total expected utility of being shown [an ad]" and "generate[s] bids for the upcoming advertising cycle," charging more or less based on the expected likelihood of audience "action," of making a purchase.[112]

We also get a sense of how Intersection uses location data by considering what Titan and Control Group had been up to prior to Intersection, CityBridge, and their bid for the payphone contract. Both companies toyed with data capture to support their out-of-home advertising ambitions. Titan found itself in hot water after the company was caught placing Bluetooth Low Energy (BLE) beacons on its payphone stock without notifying the public.[113] Beacons send and receive device identifiers that can be logged to generate extensive records of users' location data without consent, much like LinkNYC's capabilities with Wi-Fi and MAC addresses.[114] And while Titan agreed to remove the beacons after *BuzzFeed* reporters discovered their presence, the same technology is now installed on the Links' sensor panels.[115]

Control Group was more restrained in its actions but equally interested in data capture. In 2013, Colin O'Donnell, then Chief Technical Officer of Control Group and now Chief Innovation Officer at Intersection, registered U.S. Patent 2013/0107732 A1 for an invention called "Web-Level Engagement and Analytics for the Physical Space"—a combination

Wi-Fi network and digital out-of-home advertising platform that hues closely to the real-world capabilities of a system like LinkNYC.[116] The patent demonstrates explicitly how a Wi-Fi network can exploit even the most skeletal metadata to infer details about users' demographics and then leverage those inferences to target and price ads—much like the "Self-Optimizing Digital Signage" model. The patent includes dozens of potential use cases for tailoring advertising on public displays, including "event"-specific applications. "In one embodiment," the patent suggests, the system could be used "to determine that a user is attending a specific event":

> [A] WIFi access node at a sport venue or concert venue may sense that a user's mobile device is present (by sensing the device's MAC address). Upon sensing the presence of the device at a specific time, the WIFi access node can access a schedule via the external network in order to determine whether there is a sport or concert event at that specific time. If there is a scheduled event, the WIFi access node can provide sports or concert related information to the user directly via the user's mobile device or via a public channel in the presence of the user.

This kind of inference may seem rudimentary, perhaps even benign. If a user is sensed at an event, presenting information to that user about similar events makes sense, in the same way that Amazon or Netflix makes recommendations based on a user's previous activity. Things get more complicated, however, when the system starts to predict a user's identity based on this information. As the patent continues:

> In addition, based on the fact that a user is present at a sport or concert event or located in a specific location, the WIFi access node may be able to correlate additional demographic information with the user and store the demographic information in a profile related to the user. For example . . . attendance at a specific rock concert may identify the user as a male, urban, thirty–forty year old in a certain income level, or identifying a morning commute which starts in a certain neighborhood may identify the user in a certain demographic. Based on the additional demographic information, the WIFi access node may provide related content to a user.

The patent specifies the methods needed for all this inferential work, translating from "users looking at the display" to their "identity" by keeping extensive records of time and location data associated with the users. The patent goes further still, detailing methods to identify when two devices belong to coworkers, friends, or romantic partners; when groups of devices share a "common denominator" in their preference for, say, food, fashion, or football; or when a device owner is a repeat customer at a store or market. Per the patent, all these details can be inferred from only time and location data—from "events" tracked over the network's life span:

> If, over a period of time, the WIFi access node repeatedly senses the presence of a certain mobile user device (e.g., once a day), the WIFi access node is able to determine that the user with the reoccurring mobile device is a returning visitor. In response to sensing a reoccurring visitor (i.e., a reoccurring MAC address), specific content related to the fact that the visitor is a reoccurring customer may be displayed on a public channel close to the user. For example, upon sensing the arrival in a store of a reoccurring customer, a more familiar welcome message (e.g., "Great to see you again!") may be displayed on the public channel closest to the customer. Also, based on the timestamp metrics related to when the reoccurring customer usually arrives (e.g., saved in the cache of the WIFi access node or in a profile in the database), the WIFi access node may be able to determine, based on a substantially different timestamp associated with the mobile device when the user arrives, that the reoccurring user is running late and display a message ("Running late today?") on the public channel.[117]

As these and other claims illustrate in obsessive detail, with the right calculative infrastructure and responsive architecture, even "technical," non-"personally identifiable" information collected by the Links could bring O'Donnell's "Web-Level Engagement and Analytics" system to life in the physical spaces of New York City's streets.

Intersection's Dave Etherington celebrates this very potential. He sees in LinkNYC abundant opportunities for experimental cross-platform, cross-brand partnerships. As he opined on the *Behind the Numbers* podcast, "We can take some interesting third-party data, like weather data, and overlay that onto [the LinkNYC data]. There's a host of online advertising that's really predicated on these types of inputs—like, the weather gets to a certain level and then it becomes a great opportunity to order an Uber or something like that."[118] And Etherington's company has been busy seeking out this type of partnership. The week after *Avengers: Infinity War* was released, for example, Intersection partnered with Disney and Marvel in a "first of its kind out-of-home campaign" that displayed movie posters alongside showtime information at the nearest theater—"Playing at AMC 84th Street 6 in 19 min."[119] As Etherington sees it, these and other creative collaborations place Intersection at the bleeding edge of marketing's frontiers, "the future of physical advertising": "We're building the type of advertising network that belongs in a responsive city, that responds to the world around it."[120]

In Intersection's vision of the "responsive city," urban media react to, even anticipate, consumer behavior—"nudging for profit." This is precisely what legal scholar Ryan Calo has described as "digital market manipulation," media that capitalize on consumer irrationality through personalization and discrimination.[121] Although "responsive" or "interactive" in appearance, the point of such manipulations is to facilitate what communications scholars Joseph Turow and Nick Couldry describe as "the opportunity to record individual audiences' behavior *with media*."[122] Marketers do not want feedback—to know what audiences think, at least not in any meaningful sense. They want to know what audiences *do,* and particularly what they do before, during, and after seeing an ad. The same is true for Intersection, which is all but guaranteed to use the Link data to automate and optimize the trade in "audience commodities"—the buying and selling of market segments and viewer blocks.[123] Intersection's clients will not purchase screen time; they will buy access to blocks of urban audiences, categorized into neat demographic profiles of age, race, income, geography, religion, sports loyalties, and so on.[124]

And while such audience commodification will appeal to Intersection's investors (namely Alphabet), it portends a much darker horizon for civil liberties and privacy advocates, some of whom have been deploying creative campaigns to call attention to the kiosks' surveillance. Someone, for instance, has been placing smiley-face stickers over the Links' cameras.[125]

The activist group Rethink LinkNYC produces short films and regularly stages performances in public parks calling out the network's "spying."[126] In one campaign, Rethink LinkNYC activists likened LinkNYC's video feeds to the New York Police Department's notorious unwarranted search and seizure campaign—"digital stop and frisk" (see Figure 6). To these and other activists, LinkNYC is little more than "an opportunity for surveillance, data collection and corporate profit boxed and marketed as free WiFi." In its place, community organizers are demanding alternatives—communally managed Wi-Fi networks and broadband cooperatives, for example, with local organizations controlling what information gets collected, where it is stored, and for how long.[127] In a remarkable sign of the threat that these alternatives pose to corporate hegemony, twenty-six U.S. states have already passed legislation outlawing communally owned broadband.[128]

Figure 6. The activist group Rethink LinkNYC raises awareness of LinkNYC's corporate surveillance through media campaigns. The flyer shown here characterizes the Wi-Fi network as "Digital Stop and Frisk," in reference to the Bloomberg administration's unconstitutional policing strategy of warrantless stops, searches, and seizures that disproportionately affected racial-minority communities. Photograph courtesy of Eric Purcell, Flickr handle @ep_jhu (https://www.flickr.com/photos/ep_jhu/47795352711/in/pool-gothamist/).

A Public Benefit?

Public–private partnerships like LinkNYC have become the de facto model for smart city financing—a "future in which brilliant innovation is funded by smart advertising."[129] Big Four consulting firm Deloitte, for example, highlights DoITT's agreement with CityBridge in its *Funding and Financing Smart Cities* series, explaining that "the city provides concessions to allow the consortium to install the kiosks (at no cost to taxpayers) and collect advertising revenue, which is shared with the city at an agreed-upon rate."[130] Of course, all of this sounds good. In theory, public–private partnerships are win–win scenarios: the government grants the authority to oversee public infrastructures to private entities, who promise to bring their efficiency and expertise from the market to the management of collective goods. In exchange for this authority, the private entity is expected to ensure an equitable distribution of benefits. In other words, they are expected to act like a government, as a steward of the public trust.

LinkNYC fails this test. During the first several months after the network went live, the infrastructure seemed to be doing exactly what the politicians had promised: serving the city's least privileged and most in need of internet access. A reporter for *Motherboard,* for example, did a quick tour of the Links on Third and Eighth Avenues in Manhattan and found that people were indeed using the kiosks, especially their internet-browsing function. Most, however, were homeless New Yorkers, "camped out for the long haul, for hours or even days at a time, surrounded by their possessions and browsing music videos on YouTube, making phone calls, and checking Facebook . . . only two or three of them would be likely to appear on LinkNYC promotional materials (i.e., one well-dressed woman making a phone call, or one middle-aged, casually dressed tourist waiting for his phone to finish charging)."[131]

Though anecdotal, these observations lend credence to politicians' and advocates' claims that LinkNYC could bring high-speed internet access to the city's economically marginalized. At LinkNYC's ribbon-cutting, for example, Council Speaker Mark-Viverito lauded LinkNYC as "an innovative program that will provide free internet access to New Yorkers in thousands of places across the city." Brooklyn Borough President Eric

Adams proclaimed that "by expanding access to the internet from thousands of sites in our city, LinkNYC will keep New Yorkers connected." De Blasio even suggested that LinkNYC would help in "leveling the playing field, providing every New Yorker with access to the most important tool of the 21st century."[132]

Within months of the network's launch, this access would be curtailed. A slew of complaints began pouring into City Hall from residents, business owners, and local politicians lamenting "undesirable" LinkNYC "misuse."[133] Just as *Motherboard* reported, the Links were indeed attracting people; they just were not the "right kind" of people. According to the complaints, the only users were "people who linger for hours, sometimes drinking and doing drugs and, at times, boldly watching pornography on the sidewalks."[134] People were, in other words, using the internet in public space—with all its warts and hiccups.

In consultation with DoITT and the Mayor's Office, Intersection agreed to shut down the Links' browsing function by removing the Google Chrome app from the interface. Wi-Fi would still be available and users would still have the opportunity to charge their mobile phones—features that Manhattan Borough President Gale Brewer claimed were "the true benefit of the kiosks."[135] De Blasio performed an about-face and sided with the complainants, proclaiming that what was happening at the Links was a "disappointing" "pattern of abuse" that could only be solved if the City "cut off the ability to browse and get [LinkNYC] back to some of its other core functions."[136] Brewer even likened the loitering and pornography-viewing to drug dealers' use of payphones at the height of the war on drugs: shutting down browsing would be like blocking incoming calls on the payphones, a tactic that police found effective in removing the drug trade from Amsterdam Avenue.

Most complaints seemed to come from business association leaders and borough presidents, representatives whose constituents demand the type of "quality of life" assurances that drove New York City's violent and uneven policing strategies of the 1980s and '90s.[137] This was a time when the NYPD pioneered its famous "broken windows" and "order maintenance" policing styles, cracking down on low-level crimes like turnstile-jumping and fare evasion, and ramping up its "stop and frisk" campaign—all under the assumption that minor transgressions are inexorable gateways to criminality. The era that Brewer invokes so nostalgically saw the police exercise discretion to determine which behaviors—and, by extension, which communities—constituted a transgressive threat to the social order. Shutting down the browsing function, Intersection assumed a similar authority.

Although the indictments placed blame squarely at transgressors' feet, one could just as easily argue that LinkNYC's "abuse" was entirely predictable. Neither DoITT nor Intersection thought to reach out to institutions with experience in managing public internet access—namely, libraries, the traditional stewards of public information. Librarians have decades' worth of experience navigating the complexities of a public internet. As Shannon Mattern reasoned, "The information commons is messy; that's life in a robust democracy. What works in a public library can work on the street."[138] Rather than partner with experts, however, Intersection simply conceded to the demands of business leaders and cut off internet access for users who did not already own a smartphone or laptop.

Such blind spots and exclusions are common to public–private partnerships. When the public benefit of something as rich and complex as the information commons is pegged to private profit through data collection, it is easy to cast aside groups who do not contribute to the bottom line. In the end, the most plausible explanation for Intersection's decision to shutter the browsing function were the optics. If Intersection hoped to scale by selling Link networks to other cities (as the company has done since in London, Glasgow, Newark, and Philadelphia), then it needed to ensure that LinkNYC's rollout appeased key stakeholders. Any connection between the kiosks and the homeless or pornography would taint the system's reputation.

The contrast between legitimate LinkNYC "uses" and transgressive "abuses" mapped onto visual tropes of users and abusers. In promotional images (Figure 7), we see cars and pedestrians stream by a Link, blurred in a graphic representation of mobility and movement. Racially ambiguous pedestrians pass the kiosks nonchalantly, their shopping bags swinging as they walk, seemingly oblivious to the Links' presence. In this diegetic world, LinkNYC is already normalized, faded into the background of the smart urban landscape—an accepted and expected sidewalk feature, like streetlamps, mailboxes, bus shelters, or, indeed, payphones. In images of

Figure 7. A promotional rendering of LinkNYC released before the system was constructed shows pedestrians and traffic streaming by as the kiosk fades into the background of the urban landscape, like the payphones did in the twentieth century. Image by CityBridge (http://civiqsmartscapes.com/img/solutions/LinkNYC-Media-Kit.pdf).

LinkNYC "abuse," by contrast, the subjects are homeless, loiterers, pornography viewers—easily identifiable as what sociologist William H. Whyte called the "undesirables" of public space.[139] In most reporting, the "perpetrators" were nearly always Black men, even if this was not representative of actual use. And they were always stationary—on the ground, on overturned newspaper bins, milk crates, or discolored and discarded office chairs, obscuring the way for the bustle of New York City's sidewalks (see Figure 8).

In these contrasting imageries, physical mobilities intersect with representational tropes and technological practices, a complex imbrication that Tim Cresswell describes as the politics of mobility.[140] According to Cresswell, social differences articulate through differential mobilities, whether in the form of physical movement and motility, or as technologically mediated connection and access to information. In either case, one's (in)ability to move or communicate are "both productive of social relations and produced by them."[141] With LinkNYC, social distances map

Figure 8. In contrast to promotional renderings, after LinkNYC launched, press imagery focused on homeless people of color "misusing" or "abusing" the kiosks by using the browser function for extended periods of time. For example, the New York City CBS affiliate used the image shown here in a feature called "Homeless Hotspots?" describing "people making themselves comfortable at the kiosks" as a problem for New Yorkers. Photograph courtesy of Peter Michael Marino.

onto normative expectations for how technology *ought* to be used, how one *ought* to move on the sidewalk. Sanctioned use corresponds with wirelessness, fluidity, continuous flow; transgression occurs when users halt to make physical contact with the interface—a "hardline" connection of the body to the touchscreen. Like capital and information, LinkNYC's users must be in circulation; any stoppage to that flow disrupts the value chain of its ad-driven business model.

Ultimately, what de Blasio first celebrated as bridging digital divides and later decried as LinkNYC's "unintended consequences" turn out to be the same thing: people accessing the internet. Removing the browsing function but leaving the Wi-Fi allowed CityBridge and the administration to save face—to continue celebrating LinkNYC's public benefits while putting distance between themselves and any unwanted attention or negative connotations. It is useful to remember, too, that at the root of these social hierarchies are logistical mediations underlying Intersection's digital out-of-home advertising ambitions. "Undesirables" are not a commodifiable audience; they disrupt the flow of "events" that media brokers trade in.

If LinkNYC continues to serve as the model for "ad-funded smart city infrastructure," such technologically mediated reifications of social value will only become more common. In October 2017, Sidewalk Labs announced its partnership with Toronto to develop the Quayside "smart" district along twelve acres of publicly owned lakefront property. Promotional materials for the development show a serene urban life in Quayside, with residents at work and at play and all the boxes of high "livability scores" checked off. According to the planning documents, this techno-urban serenity will be propped up by what Sidewalk calls a "digital layer" of free, high-speed Wi-Fi on a network of kiosks entirely reminiscent of LinkNYC—"a single unified source of information about what is going on" that would make Quayside the "most measurable community in the world."[142] Like LinkNYC's privacy policy, Sidewalk Labs has committed to never sell *personal* data to third parties or use it to target advertising; it has even promised to inform Quayside's residents whenever their personal data is being gathered. But we have also seen how even the most skeletal of "technical" data and metadata can be marshaled to serve a profiling function. Which groups will benefit from Quayside's "responsive architectures"?[143] What activities will be deemed acceptable and appropriate, and which will its urban managers find "disappointing"? How will the "digital layer" deal with the politics of race and class that map onto the bodies of the system's users and misusers?

Revisiting Reinvention

Torontonians have advanced a grassroots resistance to Sidewalk's plans, demonstrating an acute understanding of the frictions that systems like

LinkNYC generate.[144] As LinkNYC, Quayside, and other public–private smart city partnerships proliferate, we need new models and metaphors for positive critique, for "rethinking" the ad-funded smart city (as the activist group puts it). For it is only through a cultivated sense of resignation that systems like LinkNYC come to seem inevitable.[145] Can we find inspiration for alternative designs, not by looking forward, but by looking *back*? What might a "smart city" look like if we took seriously the public's blue-sky ideas—prototypes and designs from a moment before the gray reality of procurement and franchise monopolies had set in, concepts that do not default to already-vetted features or resolve to a Faustian bargain between public benefit and advertising's logistical demands of circulation, profiling, and targeting?

In what follows, we hear from three teams that submitted prototypes to the Reinvent Payphones Challenge but which refused to conform as "variations on a well-established theme." These prototypes do not simply reinvent the payphones for a digital era. They offer a progressive counter-semiotics of the future, organized not around revenue streams or ubiquitous data collection, but around community, access, and expression.

NYC Via

Consider first NYC Via.[146] Rather than replacing New York's payphones with a new communications system, NYC Via proposed a series of networked hubs—physical shelters like bus stops—equipped with software to smooth out the seams of New York's various transportation networks.

The concept was the product of a collaboration between technology writer Meredith Popolo and designer Alex Levin. The two came together with complementary skill sets: Meredith with her background in technology, and Alex skilled in design and programming. At the time, Meredith worked as an editor for *PC Magazine* and received a press release for the Reinvent Payphones Challenge. She became interested immediately and contacted Alex, her "go-to creative friend," who had just opened his own agency. Meredith and Alex started kicking ideas around right away. It was December 2012. App-based ride-hailing services like Uber and Lyft had only recently launched in New York, and Hurricane Sandy was fresh on everyone's mind. In fact, Sandy influenced NYC Via's design in a couple

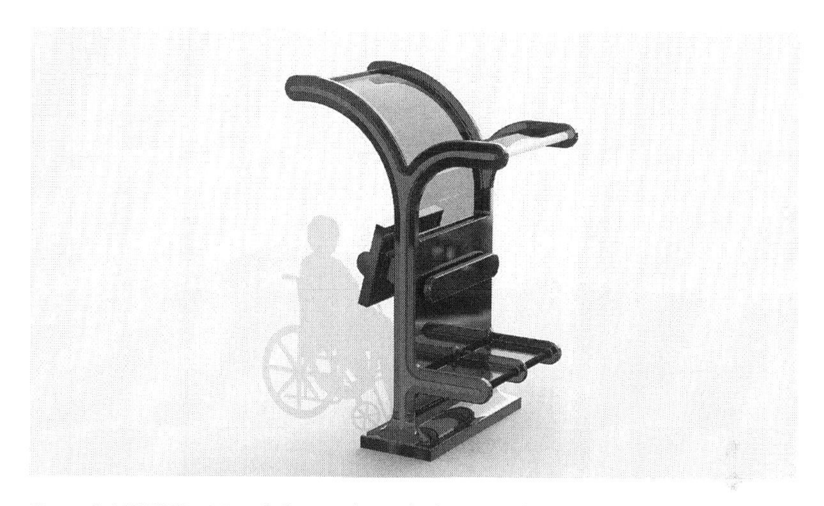

Figure 9. NYC Via, Meredith Popolo and Alex Levin's entry to the Reinvent Payphones Challenge, would create small transportation hubs linked by software to all of New York City's public transit networks. The concept also included a fare-splitting feature for the city's yellow cabs at a time before on-demand ride-hailing services like Lyft and Uber were common. Image courtesy of Alex Levin and Meredith Popolo.

of ways. For one, the designers saw no reason to get rid of the payphones' landline connections. During the storm, most of the cellular networks and internet connections had gone down as a result of cascading power outages. Landlines, for those who still had them, were essential.[147] As Alex told me, "I remember there was a huge issue with Sandy, just with connectivity, and we saw a big opportunity there—in the fact that there were all these landlines already in place and just keeping them around rather than making them totally dependent on the internet's beck and call, or relying on the [main] electricity grid." Instead of working to expand "connectivity" (which they assumed other entries would), NYC Via would take advantage of existing infrastructure. To Meredith, Wi-Fi in particular just was not that interesting as a concept. "I think we kind of figured that most people would make it like a Wi-Fi hotspot and, to me, that's less exciting. Because with [Google's New York City headquarters] right there in Chelsea, I kind of predicted that they would have free Wi-Fi over the whole city in a matter of years. So, it's like, who really needs another Wi-Fi hotspot?"

Sandy also got them thinking about transportation. The loss of electricity, cell phone connections, and, for many in the city, heating, was compounded by the closing of New York's vast subway system, which physically isolated millions who depend on public transit.[148] Was there a way to use software and other technologies to enhance the utility of existing transportation infrastructure? The challenge they created for themselves was to combine the physical features of something like a bus shelter with the resilience of payphones, but better networked and with good software—a one-stop spot for accessing all of the available transport options in the city, public and private.[149] With NYC Via, the physical structures would be hardwired and networked by landline, with a design intended for reliability, locatability, and accessibility—for instance, with a layout suitable to a spectrum of able-bodiedness and software available in multiple languages, no smartphone required.[150]

Although the concept was inspired by bus shelters and ride-hailing platforms like Uber, it also departed from them in compelling ways. Unlike bus shelters, Meredith and Alex wanted NYC Via to have an elegant and minimal structure, to take up as little space as possible on New York City's already-crowded sidewalks. But they also wanted the kiosks to be a place where people could stop to catch their breath. Unlike LinkNYC's mobility politics, the shelters would allow pedestrians to slow down, to pause before getting on the move again.

Alex and Meredith also wanted the designs to be as "iconic" and "distinctive" as the city's yellow cabs. For the structures to be successful, they needed to be easily identifiable. As Meredith put it, the idea was for someone "to look down the street and be like, 'That's a place where I can get a ride.'" They conceded that the structures' iconic profile would be attractive to marketers. They were not opposed to advertising—in fact, their submission included a mock-up structure wrapped with a design by the stylish Finnish home goods store Marimekko. But they also stressed that the ads would have to cohere to the structures' iconicity, not the other way around. The design would not accommodate marketing; marketing would have to accommodate the design. And they were not much interested in data-collection unless it served transit-related objectives.

Where NYC Via drew on ride-hailing apps like Uber, it did so in ways that capitalized on the ingenuity of networked communications while avoiding the pitfalls. As I will discuss in greater detail in the next chapter, platforms like Uber and Lyft capture economic value by shirking employment regulations and outsourcing overhead costs onto workers. NYC Via's transportation aggregator, by contrast, would merely connect users to existing transportation options. Since the system would be embedded in specific locations, all users would need to do is enter their destination and they could access all nearby transportation options—bus, subway, rail, taxi, airport shuttle, and so on. The submission even included a concept for NYC Via's native software, which they called FareShare, to encourage opt-in cab-sharing.

Rather than harping on connectivity, then, NYC Via presented an opportunity to network together the best of what New York City's complex transportation networks already offered. While it may now be common to see this type of aggregation on apps like Google Maps, many of those programs quietly harvest and sell users' location data to the highest bidders—analytics and data brokerage firms that traffic in audience profiling and targeting.[151] NYC Via, by contrast, was envisioned as a public utility, a service that the City could maintain without a profit motive, a refuge in what can be an overwhelming experience of trying to catch a cab or figure out the bus schedule. As Alex put it, "We never had the idea of it being just the internet. It was always going to be a way to access local places, like to see what stuff is going on nearby, and to provide just a dedicated space—ideally, say, one [parking] spot in front of the actual phone booth, where a cab would actually be picking you up." It is easy to imagine such a system immediately opening up New York's transportation options to thousands of new users—for instance, subway riders unfamiliar with the bus system, or tourists overwhelmed by the subway's complexity. NYC Via operationalized a vision for the smart city that was about improving on infrastructures already in place—a way to open up access to existing services, not "just the internet."

Windchimes

Like NYC Via, Windchimes set out to repurpose the payphones' existing landline infrastructure.[152] But unlike NYC Via and nearly all the other entries to the design competition, Windchimes did not propose to replace the payphones with anything. Instead, it would simply augment the existing phone stock with a suite of low-cost sensors for environmental monitoring and local activism. The approach caught the attention of the competition judges, earning Windchimes a spot as a finalist in the "community impact" category—the only entry to bring a working prototype to the demo day, and the only submission by students, not professional architects or designers.

The Windchimes concept was developed by a relatively large team, but at its core was a collaboration between Nick Wong and Ann Chen. At the time, Ann was in the process of completing a master's degree in the Interactive Telecommunications Program (ITP) at New York University's Tisch School for the Arts, where her research addressed environmental sensing and construction, with a particular interest in rooftop farms. Nick had been working on his master's in mechanical engineering at Cooper Union, where he had become enamored with the challenge of repurposing analog infrastructures for the digital age. The two met at a Meetup at ITP organized for students interested in submitting to the

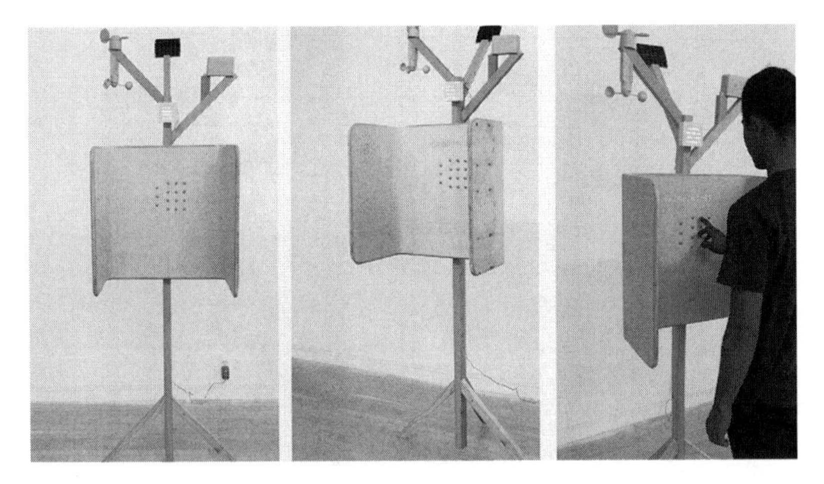

Figure 10. Windchimes was the only winning entry submitted by students, not professional designers or architects, and the only submission to bring a working prototype to the Reinvent Payphones demo day. The concept augments existing payphone infrastructure using low-cost sensing equipment to monitor environmental conditions. The submission also included a community hub for data management and administration. Windchimes by Nick Wong, Ann Chen, Rena Lee, Eric Leong, Louis Lim, and Paul Miller. Photograph courtesy of Ann Chen.

Reinvent Payphones Challenge and were joined by others graduate students in design, technology, and art.

The Windchimes concept evolved as a symbiosis of Nick and Ann's specialties—Nick with his interest in new uses for old technologies and Ann with her knowledge of environmental sensing. The team assumed that most other submissions would "go digital," and this was not something Nick found terribly interesting. He had been reading about efforts to transition residential phone services from landlines to voice over internet protocol (VoIP), a campaign that he saw as problematic and even dangerous given VoIP's susceptibility to disruption. Landlines draw electricity through copper wiring and because they are "off-grid" stay powered throughout outages.[153] Like Meredith and Alex, Nick wanted to make use of the landlines' already robust design. He therefore formulated a design challenge for himself—to use the phone lines for purposes other than end-to-end conversation. When the team worked through a new concept, Nick would ask himself, "What would this allow me to do that a smartphone wouldn't?" And as he pondered this creative constraint, he found himself turning away from ideas that involved internet connectivity or even digital displays, instead relying on the payphones' native technical affordances and geographic ubiquity. "What interested me most about the competition was the idea to *reuse* the payphones—not to replace them, but to think about new ways to adapt what you already have with a landline. Thinking beyond digital media and connectivity—we just figured no one else would be."

Eventually the team discovered an old industrial hack, exploiting the phones' touch-tone dials (0–9, #, *) to transmit data. This technique, called multifrequency signaling, had been used for decades in tasks like automatic redial, fax machines, dial-in services, and dial-up modems. And although it was Nick's idea to use the touch-tones, it was Ann's interest in environmental sensing that gave a focus to the design. Windchimes would augment the payphones by attaching a rig of low-cost sensors to capture and transmit environmental data across the phone lines to community data centers: "We imagined New York City's existing 11,000 payphones as a distributed sensor network providing real-time and hyper-local records of the city's rain levels, pollution and other environmental conditions."[154] This was a unique application of the research that Ann had been developing through her graduate work on urban agriculture and ecology. "There was a lot of thought put into the data that would be collected and what kind of sensors we wanted to put on the phone," she told me. "It could tell each neighborhood what the noise pollution levels were . . . but also a whole host of other data and could potentially even be used to broadcast events or things going on. So, it could really become a community hub, or at least that was the idea. The payphones were already so spread out throughout the city, in what was really a democratic kind of way."

While other entries also focused on environmental applications, Windchimes' elegant simplicity stood out. The rig was more rudimentary than even a dial-up modem, and the prototype could be connected to a phone line and start recording data as early as the demo day. The team had even built a series of open data portals that would give local communities access to and control over the data. The rigs would not cost much; Nick and Ann estimated that each unit could be constructed for around $100, meaning that a citywide network could be up and running for just over $1 million (whereas CityBridge estimated $300 million for just the first year of operations). The design's insistence on utility privileged functionality over form and style, practicality over novelty, public access over private benefit, and an ethic of making and doing over the visual richness of speculative design. The low-cost, modular, open, "plug and play" assembly presented a stark contrast to other finalists' smartphone-inspired black-box operating systems.

Digital Democracy

Finally, consider Digital Democracy.[155] This entry did not make it to the finalist stage, likely because the concept was too guileless. Digital Democracy proposes to simply replace the payphones with large LED displays—or "digital canvases"—onto which anyone can upload content: flyers, posters, original artwork, local business promos, yard-sale signs, and so on. The concept imagined the city as a radically open communications space, free of censorship; the system would simply display whatever content its users desired. The only constraint would be proximity. Like graffiti or flyers posted to a utility pole, users would have to upload their content by physically going to the signage—by actually being there.

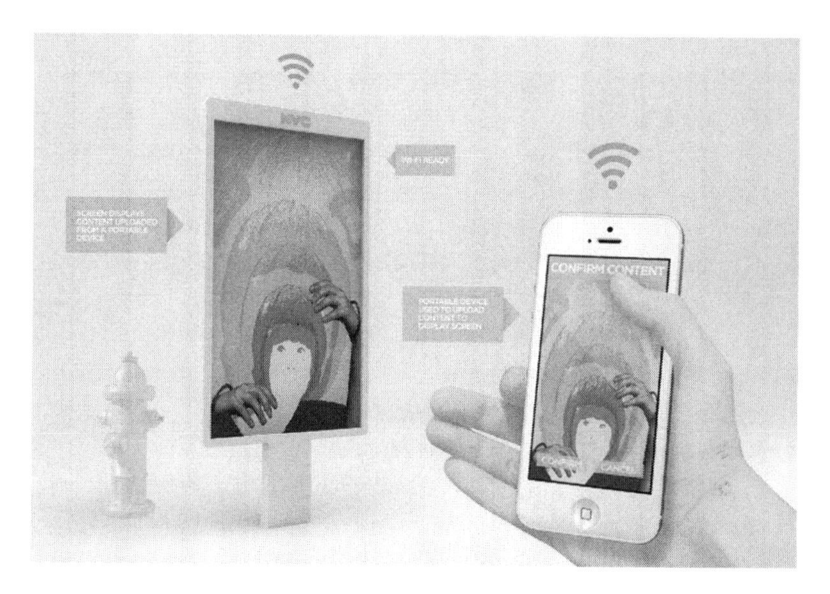

Figure 11. Inspired by graffiti and other street media, the Digital Democracy concept would have replaced the payphones with large "digital canvases" available for anyone to upload visual content. Designers Will Arnold and Andrei Juradowitch conceived of the system as an "exploration of a certain way of looking at free speech in the twenty-first century." Image courtesy of Will Arnold and Andrei Juradowitch.

And like those pre-digital street media, Digital Democracy would be ephemeral by design. An image would only stay up until the next user came along to replace it.

Digital Democracy presents as a variation on the kinds of "hyper-local urban communication" that have animated cities and urban life for millennia.[156] Designers Will Arnold and Andrei Juradowitch came up with the idea while working together at the office of a design agency. They had already been spitballing concepts for a while, looking for a project outside their day-to-day work. When they heard about the Reinvent Payphones Challenge, they tried to convince the firm to submit something. Sensing a lack of interest, they went for it themselves.

Like Windchimes and NYC Via, Will and Andrei tried to incorporate existing infrastructure to make a new system—"whether it would be the actual phone booth itself or the fact that there's electricity being pumped into the components." They were not put off by digitality in the same way that Nick and Ann were, but they wanted to think beyond the passivity that they perceived as the dominant motif in urban technology. As Andrei explained, "We wanted to create something where we gave people control . . . something where people could add to it rather than just observe it. When you look at way-finders and maps and stuff, it was all very, like, sort of observation; you didn't really engage with it. You just read information from it." Digital Democracy would be the exact opposite—a platform for citizens to express themselves, to use the city as a democratic medium. An "exploration of a certain way of looking at free speech in the twenty-first century," as Will and Andrei put it.

Advertising on the backs of payphones may have taken off in the 1990s, but Will and Andrei observed an explosion of advertising in public spaces in the years leading up to the Reinvent Payphones Challenge. They envisioned Digital Democracy as a counter to the corporate takeover of the visual environment, "just giv[ing] a small corner of the city or some little piece of real estate back to the people—to advertise whatever was on their mind—as opposed to these really expensive corporate advertisements on the side of a phone booth. It was kind of giving a slice of the city back to the people so they could put meaningful images all over it." They spoke of graffiti as an inspiration: "For decades, residents of New York have been relying on public space and private property as outlets for personal expression and artistic exploration. As a result, every corner of the city contains some sort of graffiti, sticker art, flyers, or wheat-pasted posters."[157]

Digital Democracy co-opts certain affordances of digitality—the ability to "slide in" content from a smartphone or tablet onto a public display—but only in service to a space dedicated to the time-honored, do-it-yourself irreverence of street media.[158] The goal was to increase the range of communicative forums available to activists, artists, and community organizers without having to rely on corporate sponsors: "We felt like New York was maybe one of the only cities where this could actually work. It'd be a utility for the city. It seems really simple and totally achievable, but in reality, really hard to imagine existing. If it did, the question would become, what will people use them for? I think you'd start to get really interesting answers."

It is doubtful that we will ever get to see those answers. Intersection has launched programs to get local artists' work up on LinkNYC screens, but these are far from the direct public-access model that Digital Democracy imagined. In one variation, Intersection employees scour Instagram for "influential" local artists, and then select works to be rotated on the Links' ad panels.[159] If you cannot win this lottery, artists have the opportunity to pay a $75 fee to enter a competition and, if selected, must cough up an additional $875 in exhibition costs to display their work on a Link.[160] Likewise, Intersection offers a special ad-buy program for local businesses, but the minimum rate of $5,000 for four weeks of ad space (with prices jumping rather quickly for busier locations and display times) makes it unlikely that many small-business owners will be able to afford the service.[161] In Digital Democracy's radical vision, fees are not an issue. The network would be a public communications utility—not a prime-time commercial slot or a banner ad, but a community bulletin board, a public access station.

Logistics and the Design Imaginary

LinkNYC replaced New York City's aging payphones, an infrastructure whose disrepair and neglect were the products of a changing technological landscape and shifting political–economic priorities among operators and urban managers. The new network embodies the city's entrepreneurial turn away from communicative utility to commercial value, imaginable only through partnerships with corporate capital hungry for data. While the old payphones always involved some inherent profitability (after all, they were never free), they nonetheless represented a straightforwardness and neutrality that are fast becoming scarce in the ad-funded smart city's public–private infrastructures. Are we sure these are qualities that we are ready to forego as we upgrade our cities' systems and services?

Such questions may seem like they are beyond the reach of everyday citizens. But in the end, it is our collective imaginary that shapes the design of our cities' infrastructures, and those designs in turn that mediate citizens' access to urban infrastructure.[162] This dialectic helps explain how New Yorkers ended up with an infrastructure like LinkNYC—a system that offloads citizens' data in exchange for advertising revenue; a system whose legitimacy derives from its supposed inclusivity but which reneges on its promise of public information access. At each point along the network's trajectory, we witness uneven struggles to define its features, to police appropriate and inappropriate uses, to render certain operations visible while shrouding others. While the city's contract with CityBridge all but ensures LinkNYC's reign for the remainder of the franchise agreement, those moments of struggle and contestation reveal cracks in the smart city's interface, openings for alternative visions—not just of "the city of tomorrow," but for the city of today. What kind of social and technical worlds do we want to build for ourselves, for our communities?

If we leave these questions in the hands of technology firms, the answers we will get are predictable: their primary concern is and will always be with the bottom line. Writing for *Smart Cities Dive,* for example, Michael Provenzano, CEO and founder of marketing and analytics firm Vistar Media, explains that Alphabet's investments in Intersection and LinkNYC via Sidewalk Labs are hardly happenstance. As early as 2006, Google had expressed interest in using Wi-Fi to support an out-of-home "contextual" advertising network in San Francisco.[163] Today, that interest is now an imperative. Nearly 90 percent of the $90 billion or so that Alphabet brought home in revenue for 2016 came from Google's online advertising sales. Ads finance all of Alphabet's "moonshot" projects, from self-driving cars to life-extension biotech to Quayside. But while Google is selling more ads than ever before, the amount it brings in for each ad (or the "cost per click") is dropping.[164] To counter this trend, Provenzano argues, Alphabet needs to expand and innovate its ad business, and out-of-home marketing is the most viable option:

> As cheap prices, ad blocking software and savvy consumers drive [Google's] online ad business down, [Alphabet] must look out the window to the outside, physical world—where it's impossible for us to ignore messages, and where traditional media is rapidly evolving thanks to new technology and connected devices. Alphabet, in other words, needs to double down on its urban innovation business.[165]

And double down on urban innovation is exactly what Alphabet did—a "spatial fix" to its crisis of shrinking revenue streams.[166] Citing a recent report from consulting firm PricewaterhouseCoopers, Provenzano predicts that digital out-of-home ad sales will explode in the near future, in part thanks to Alphabet's investments in location-based data: "Location-based data will allow advertisers to reach their audiences better than ever, which means the value of each ad will increase. That means Alphabet will be able to say goodbye to declining costs per click." LinkNYC is the vanguard of these efforts, Alphabet's entrée to the digital out-of-home gold rush.[167]

Two interconnected points are worth emphasizing. First, there seems to be an emerging consensus among smart city practitioners that digital out-of-home is the best way to finance smart city infrastructures. In Provenzano's admittedly biased view as CEO of a marketing firm, "revenue from out-of-home advertising can provide the seed capital needed to subsidize the initial cost of building the connected cities infrastructure." But Provenzano is not alone in imagining this "future in which brilliant innovation is funded by smart advertising."[168] Big Four consultant Deloitte cites LinkNYC as an exemplar of public–private success. Funding models based on advertising revenue, the Deloitte report explains, "repurpose existing physical assets, bring in private-sector capital and expertise, and ultimately create new sources of revenue through data collection and citizen engagement."[169] Nor are connections between advertising and smart city financing lost on advertising trade associations. The Digital Place-Based Advertising Association is quite pleased with the new partnership models, quoting Brian Dusho (a "foremost expert in the smart cities space") to define smart cities as "out-of-home [advertising] meeting the Internet of Things on the street."[170] In these and other visions of the future city, advertising is the only conceivable mechanism for enlisting private capital, which, in turn, becomes the only conceivable funding source for urban infrastructure.

But this future is a two-way street. If urban infrastructure is beholden to ad-driven financing, advertising is equally dependent on the city for its data-crunching machinery. Market projections expect digital out-of-home to remain the fastest growing industry sector well into the 2020s. Allied Market Research projects a jump in digital out-of-home revenue from about $3.5 billion in 2016 to over $8 billion by 2023, with a growth rate of 12 percent per year.[171] Other projections are even more bullish, projecting the industry at upward of $32 billion by 2025.[172] While industry insiders might quibble over these discrepant projections, they agree that this growth will be pegged to the expansion of the industry's already insatiable appetite for data, and that expansion means moving into physical spaces. A report from the Interactive Advertising Bureau and PricewaterhouseCoopers insists that out-of-home advertising's future will not merely be digital; it will be *programmatic*, with media buyers, agencies, network operators, and adtech firms embracing digitization's "automation of media buying by programmatic trading."[173] Programmatic digital out-of-home marketing (which industry practitioners abbreviate unironically as pDOOH) will require a constant stream of audience data as grist for the marketing machine, further entrenching the logistics of advertising's computational turn deep within the furnishings of everyday urban life.

What are we to make of this entanglement of the smart city and advertising? What lessons does LinkNYC offer for critics of the corporate takeover of urban infrastructure?

The mass media of the twentieth century taught us to be skeptical of advertising's ideological content. Gender and racial norms were encoded in messages, as were the cultural foundations of consumer capitalism.[174] These issues persist, but their hand may be overplayed. Critiquing the ideological content of messaging will not get us very far if we wish to tackle the conditions supporting the ad-funded smart city's never-ending pursuit of increasingly minute and increasingly fleeting sources of economic value. This is the logistical question that LinkNYC raises as it translates from the infrastructures of online advertising to the city's streets. As critical media historian Lee McGuigan argues, programmatic advertising has always been about the logistics of audience commodification.[175] Like the "integrated supply chain," programmatic advertising connects the distribution of content to its sales, the transmission of messages to the messages themselves. And through these logistical "intersections," the network produces *events*.

Events are advertising's sine qua non, its essential condition, the raw material of marketing's value. Events embed messages in social contexts. They "link" commodities to audiences' desires. And the more immediately they do so, the better. We see this, for instance, when Intersection's executives discuss LinkNYC's ads becoming tied to real-time inputs: "Whilst we don't ever look at web traffic for people on the Wi-Fi, we do know, for example, the types of devices that people have. So one use-case might be that we see lots of fitness trackers in an area and that might lead us to change an ad based upon that insight."[176] The event is a marketing opportunity, a temporary infrastructure of desire and capitalization. From an investor's point of view, LinkNYC is valuable because it *optimizes and streamlines the production of events,* pioneering a logistics of instantaneous market-making on nearly every corner of the city—as Provenzano quips, "where it's impossible for us to ignore messages."

This market-making depends on a calculative infrastructure and responsive architecture capable of detecting and responding to events wherever and whenever they might occur. Certainly, we might appreciate that the Links still offer free Wi-Fi, phone-charging, way-finding, and 311 information—all services that tourists and community members alike will surely find useful. But this convenience belies the system's business model and its techniques for monetizing ad distribution—inferring consumer profiles, targeting ads, counting "impressions," tracking movement. From this perspective, LinkNYC looks less like the public utility that politicians celebrated than a corporate infrastructure for capturing data on urban life—measuring effects, capturing attention, producing events, commoditizing audiences.

If pDOOH is the future of advertising, we can only expect to see more of its invasive logistics penetrating the smart city. That is why it is more important than ever to study alternatives. By excavating designs from a moment when the forms and forms of action imaginable in the smart city were far more fluid and open, we can begin to grasp the political stakes of technology design. Who gets to define the "efficient," how is it operationalized and optimized, and for whom? In prototypes like NYC Via, Windchimes, and Digital Democracy, we glimpse *another* smart city, another set of logistical efficiencies, where problems stem from entirely different conceptions of urban "optimization." NYC Via envisions efficiency as connection-building—linking subway and bus lines, helping taxi riders split a fare, and the like. Windchimes forsakes expensive overhaul for maintenance and functionality, enacting efficiency in the form of low-cost, modular sensing equipment that can be controlled and managed directly by communities. And with Digital Democracy, we get a system whose radical openness transforms public surfaces into a platform for creativity and voice. To these designers, engineers, and programmers, efficiency serves other ends—maintenance, care, and the repair of our existing systems.[177] They are about access, not to deals at the nearest restaurant or movie theater, but to infrastructure and expression. With LinkNYC's exclusions and covert data collection, these prototypes take on a critical edge—"testimonials to what could be," "alternatives that highlight weaknesses within existing normality."[178] That they have been buried under the weight of corporate visions for the smart city suggests a failure—not only *of* the imagination, but *for* it.

2

Control

Calculating the On-Demand Worker

> What we like to say is that the vision for Uber is the cross between lifestyle and logistics. We are all used to seeing that on the internet. With the click of a mouse we are bringing that experience to the real world. I can now push a button to get what I want and it is delivered to me.
>
> —TRAVIS KALANICK, founder and former CEO, Uber

Life in the "Uber City"

It's a Thursday night, 8:05 p.m. The push for dinner orders is over and I find myself weighing whether I want to call it quits. I've already been logged on for over three hours—delivering food by bike, shuttling from a set of upscale and fast-casual restaurants to customers' high-rise apartments in Philadelphia's wealthy Center City—but most of that time was spent waiting around. I'm standing in a small park taking field notes on my iPhone. Other than seeking cover from the rain, the most memorable event of the night was when a restaurant screwed up my order; the customer didn't seem to notice or mind that I was late, but I made the same amount that I would have without having to wait around—about $13 over the past two hours.

I didn't plan on working today, and so far, I'd only netted about $40. That light yield made working in the cold and wet seem like it wasn't really worth it—especially because I'd earned a lot more during the same slot when it wasn't cold or wet. I only decided to log on after my phone vibrated with an alert from the Caviar app. The company wanted more couriers—

probably a dispatcher in San Francisco seeing the weather in Philly and anticipating a bump in order volume. Once I started riding, though, I stayed on, figuring it would get busy. And it did, for about an hour. Right away I got three orders in a row. It's good when it's busy—you keep moving and time passes quickly. But that was over two hours ago. Now, I'm not getting "priority"—I didn't "pre-schedule" myself through the app, so if there are any orders, they're going to couriers who signed up for the shift a week ago. So much for "work whenever you want."

My socks are nearly soaked through. It's been over 60 minutes since I finished the last order, which means that I'll make less than $9 for the hour—and that's before taxes, which, as a 1099 worker, I'll have to do myself and pay at a higher rate than W-2 employees. But I stay logged on—I'm already out here—hoping for one last delivery in the direction of my house. Another courier rides by, not the bike-messenger-type with tattoos and a fixed-gear and a tiny hat; just a guy on a mountain bike, probably a college student, wearing the same bright orange, branded box-shaped thermal backpack that Caviar gives its couriers and that I, too, have on. We exchange nods. Moments later a sedan with a pink mustache on the grill pulls over to let out a Lyft customer. A man on an electric bike navigates around the car, two plastic bags full of takeout balanced on the handlebars, on his way to a delivery for a Chinese food restaurant. My phone never vibrates. I log out, call it a night, and ride home.[1]

What does life in the "Uber City" look like? According to Carlo Ratti and Matthew Claudel of MIT's SENSEable City Lab, the Uber City is an innovation utopia—a fantasy in which government prioritizes "disruption" over all else as policymakers learn from the "failures" of overregulation. To illustrate the point, Ratti and Claudel give the cautionary tale of Minitel, the proto-internet of the 1980s and '90s developed and administered by France's postal service and national telecommunications operator. Although a success in its time, Minitel's "top-down" network structure and "rigid architecture and proprietary protocols" were its undoing—the state-run experiment in information and communication was eventually overtaken by the commercial internet of the dot-com era. The moral of the story? Cities should be governed more like the World Wide Web than like Minitel: they should be agile, built from the bottom-up, neither rigid nor proprietary; and policymakers must learn that their job is not dictating how technology should work or how people should communicate, but simply "producing and nurturing the regulatory frameworks that allow innovations to thrive."[2]

That all sounds good. After all, terms like "bottom-up" and "innovation" evince an aura of citizen-driven progress. Just as the Übermensch protagonists of Ayn Rand's novels stood for individualism, self-actualization, and entrepreneurialism, the "Uber City" is about creating a space where innovation can flourish, where makers and doers can thrive in a competitive world.[3] Travis Kalanick, founder and former CEO of leading ride-hailing app Uber (from which the Uber City takes its name), certainly seemed to embody those qualities when his star was rising. But outside of Ratti and Claudel's fable and Kalanick's celebrity, not everyone gets to be an Übermensch. To any of the millions of drivers, couriers, domestic laborers, task-completers, warehouse-pickers, and data-enterers whose lives and livelihoods are now dictated by the whims of algorithmic management,[4] life in the Uber City is not about innovation at all. It is about working dogged hours to make enough money to scrape by, getting paid piecemeal for discrete tasks, and missing out on most of the hard-won workplace protections fought for by organizers and unions over the past century.[5]

If Uber is about connecting lifestyle to logistics, as Kalanick suggests, then your perspective on the Uber City will depend entirely on which side of the equation you find yourself—the lifestyle or the logistics.[6] The above vignette offers a sense of what this *other* Uber City looks like—isolating, detached, unpredictable, perhaps a bit petty. The scene, reconstructed from field notes, was common. At the end of each shift that I worked as an independent contractor and bike courier for the San Francisco–based food-delivery startup Caviar, I reflected on what happened, how I felt, where I delivered food, and to whom. Over time, I started to feel increasingly frustrated. Customers became too demanding or impatient, restaurants too slow or unhospitable, security guards too bureaucratic. For the most part, though, customers, restaurant workers, and security guards did not change—the *app* did. Caviar constantly tweaked and adjusted its interface and payment algorithm. The company introduced new, competitive challenges to win swag or earn a bonus, almost on a weekly basis.[7] They changed how information was presented, how nudges and notifications were sent, how workers communicated with dispatchers to report problems or emergencies.

The platform, in this sense, was "flexible"—constantly mutating—but the work was not. The work, like most work, was rigid and proprietary. As an independent contractor, I may have technically been free to decide when to log on or which orders to accept. But all these decisions ultimately boiled down to a single question: *Is it worth it?* Is the payout worth the hassle of picking up from a restaurant that I know to be slow? Is it worth rejecting a job if the system might put me at the bottom of the queue? Is it worth going out in the rain or snow—when order volume is up but the fatigue and risk of injury greater? Is it worth it to stay logged in for one more job, even if that means waiting without pay—sometimes for up to an hour?

The prior chapter explored how interface design can both obscure and facilitate a logistics of digital advertising in public spaces. This chapter considers interfaces of a different sort. Where LinkNYC illustrates the hazards of pegging public goods to private profits, platform work highlights the frictions of technology's integration within the city's economic circuitry.

The story is set at a moment when the city has become a key site of capital accumulation, not just with the commoditization of urban audiences (as we saw in the previous chapter) but through the intensification of mobile services—a process that urban theorist Ugo Rossi describes as the "appropriation of the urban commonwealth."[8] What role do new technologies play in this appropriation? On the one hand, workers only access platforms like Uber and Caviar through the apps they download to their smartphones. These serve as an interface of meticulous managerial control over workers' access to market information. On the other hand, if we zoom out, it also becomes clear that apps are only part of the story. Beyond the software is an interface of the conflicting realities that workers must navigate on a daily basis, where the abstract info-spaces of managerial calculation confront the lived experience of the city and all its uneven surfaces—traffic jams, potholes, high-rise apartments.[9] To make it in the smart city, workers must straddle these worlds, often thanklessly and without much support.

Platform work tells the story of the smart city and "the future of work" from below.[10] It shows how promises of innovation and entrepreneurialism can be warped and twisted beyond recognition. When I worked as a bike courier for Caviar, I was at the beck and call of the platform app. I hustled to pick up food orders and get them to the customers on time. I logged on when the platform requested more workers. Certainly, I could choose to log off, and I could decline any order that I did not want to take. I could even choose to not log on at all. But those were all decisions not to earn.

"On-demand" service platforms first exploded onto urban markets in the early 2010s—first Uber, then the others: Lyft, Postmates, Instacart, Caviar, TaskRabbit, Handy, Luxe, and so on.[11] At the time, the startups seemed to offer an entirely new way to work and consume, to get around, to eat and get stuff done. They promised to rewrite the social contract of everyday life.[12] As the narrative went, the platforms would prioritize "sharing." They would use new technologies to tap into the countless resources and time spent idle as we go about our day-to-day lives. They would connect people looking to make extra cash with folks trying to get something done or rent property. Earn money by giving someone a ride on your commute home from work! Pick up some extra cash by making deliveries while you ride your bike! Going on vacation? Rent your home while you are away! Good at putting Ikea furniture together? Get paid for it! All of this of course seemed a noble proposition. Because all that idle capacity, all those hours spent driving to and from work, all those empty apartments, would simply go to waste otherwise. Idleness is the antithesis of efficiency—unproductive and out of circulation—and "platform urbanism" would be all about making people and things productive, bringing them into the fold of economic circulation.[13]

But who gets to enjoy the benefits of this new contract? As a courier, I learned that life on the logistical side of the Uber City has less to do with innovation than with how technology reconfigures the "circuitry" of labor—the social and institutional norms and obligations that shape our work lives and livelihoods.[14] Although each platform may specialize in a particular service—Uber and Lyft do ride-hailing, Caviar and Postmates do restaurant deliveries, Instacart does grocery deliveries, and on and on—they share some fundamental features. Almost as a rule, the platforms classify their workers as *independent contractors*, not employees.[15] As a consequence—and again, almost as a rule—workers have to supply their own equipment, whether that means computers for coding, smartphones to receive a job on the go, or a car to pick someone up in. It also means that workers do not get to enjoy most of the protections of traditional employment. In fact, a good chunk of the platform workforce has to supplement meager earnings with government benefits.[16]

Of course, none of this is particularly novel for the service industry, a category that all but perfectly overlaps with the "working poor."[17] Nor do these conditions come close to the Uber City narrative of technological innovation and flexible employment. The more accurate story of the on-demand economy is that a cadre of Silicon Valley startups pioneered a unique combination of technology, appropriation, and regulatory arbitrage. This particular recipe helped the firms inflate their market valuations by scaling rapidly and then hoping to make an "exit," either by selling to a larger firm or going public.[18] Platform startups adhere to Facebook CEO Mark Zuckerberg's famous motto, "Move fast and break things." With an independent-contractor workforce, the platforms could stay "agile" while still growing quickly. They did not have to onboard new employees or even establish a brick-and-mortar presence as they entered new markets: contractors could be hired online or simply downloading an app, with only minimal vetting.[19] Investors found the speed of this scalability particularly attractive, and because the startups were flush with venture capital funding early on, they could afford to undercut rivals and even pay their contractors relatively well—at least at the beginning. In some cases, venture financing kept the firms afloat for years before finally drying up, at which point the companies hiked customer prices or slashed workers' pay rates—sometimes both.[20] Even if this did not "break things" (for instance, by wiping out competition), it certainly had an impact by altering consumer expectations in ways that traditional business models cannot afford to meet. Once customers come to expect platforms' promise of instant gratification—short wait times and low surcharges—they are unlikely to return to something as rudimentary as a yellow cab.[21]

Antitrust law would also benefit the platforms by excluding contractors from collective bargaining rights. This allowed the startups to dictate unforgiving terms.[22] In this sense, the Uber City begins to look a lot like the rest of Silicon Valley, where unions have long been viewed as a barrier to innovation—"dinosaurs" where the "tech industry is an asteroid."[23] Like regulation, worker organizing is seen to be unnecessarily rigid and hierarchical. As Intel's cofounder put it back in the 1960s, "remaining non-union is essential for survival for most of our companies. If we had the work rules that unionized companies have, we'd all go out of business."[24] The same could be said quite literally of the platform economy today. Some observers estimate that workers' (mis)classification as contractors saves firms up to 30 percent on labor-related expenses. In the United States, the contractor classification frees companies from minimum wage, overtime, health insurance, and contributions to Social Security, Medicare, workers' compensation, and unemployment insurance. Without these savings, it is unlikely that the platforms would survive, at least in their current guise.[25]

But these cost savings do not come easy. To enjoy the benefits of an independent-contractor workforce, firms legally must guarantee their workers a modicum of independence from employer control. The U.S. Department of Labor and Internal Revenue Service each provide criteria for distinguishing employee and contractor relationships based on grades of control—does the employer control how the work is done, what equipment is used, the manner in which the work is completed and when, and so on? Adhering to these regulations introduces coordination problems for the platforms that traditional employment models easily avoid.[26] Without being able to tell workers where to work, when, which orders to take, what kind of car or bike to use, et cetera, the platforms risk poor service quality, a dissatisfied customer base, and, ultimately, revenue loss.

These and other competing objectives hang in a delicate logistical balance. Here more than anywhere else, then, is where technological "innovation" kicks in again—not as a testament to "the new frontiers of digital information when it inhabits physical space," as Ratti and Claudel would have it, but as a *solution to the coordination problems that the contractor classification introduces*. If technology means convenience on the lifestyle side of the equation, then on the logistical side, it means management at a distance, soft control under the guise of "uncoerced choice."[27] Pricing algorithms, order allocation algorithms, mapping and routing functions, interface design, and so on—all these tools constrain and limit workers' decision-making such that the only options available have already been vetted and limited by the platform.

The question that chronically hounds platform workers—*is this worth it?*—is not the by-product of exhaustion or wet socks. It is the crux of a logistical rationality that platform managers hope to cultivate in workers such that their decisions align with the company's interests. This is my primary contention. In order for the smart city's platform-based economy to be successful—to satisfy an increasingly impatient customer demand, to continue expanding into new markets and capturing profits—workers must "learn" the right responses to that eternal question of "the worthwhile," and adjust their decisions accordingly. Those that fail to do so are deemed "irrational," and irrational workers do not make much money.

Algorithmic control therefore is not coercive in a direct sense. It is the false choice between nothing and too little that the platforms offer in each contract for each task or work order.[28] But it is also the case that adherence to this platform-based "rationality" will only ever be partial and incomplete. On-demand labor management draws from techniques refined in the warehouses of global supply chains, but it implements those techniques in the mess of city streets and amid a deteriorating moral economy of work. The friction between the info-spaces of managerial control and the lived experience of platform work undermines the firms' managerial strategies. Building on interviews with platform workers and my own observations as a courier, this chapter explores how workers' sense of dignity and autonomy chafes at the influence that platforms pretend not to wield. Workers negotiate opaque payment algorithms and automated queues at the same time that they navigate the city's uneven social topologies. And while they may do so more or less effectively, they rarely get much support from their employers. Instead, they get "nudges," notifications, alerts—incentives to log on when the managers need them to—and radio silence otherwise. Left to their own devices, as sociologist Julia Ticona puts it,[29] workers register these conditions as a declining moral economy of work, a rewiring of work's circuity in which the benefits of "flexible employment" are hoarded almost entirely by the platforms.

Flexible Employment?

At the end of 2015, I had been working for Caviar for over fifteen months. So when I heard linguist Geoffrey Nunberg on NPR's *Fresh Air* nominate "gig" for word of the year, my ears perked up.[30] Tracking its origins as musician slang for a performance to its later adoption by beat writer Jack Kerouac and other 1950s hipsters, Nunberg found that "gigging" had come to signify "any temporary job or stint"—the kind of work that you *do* but has nothing to do with who you *are*. That the gig economy's prominence now demanded lexical recognition reflected my own experience and observations. Platform ride-hailing had arrived in Philly three years earlier when Uber entered the market, with Lyft launching not long after. By the end of 2015, both were still operating illegally (and continued to do so through October 2016) but nonetheless growing rapidly.

My brother started delivering for Caviar in the spring of 2014 and I followed suit a few months later. We made good money in those early days. On Caviar, couriers earned a ten-dollar-per-order flat fee, which meant that if you completed two orders in an hour (not much of a hustle), you would do well. These were the "golden days" of platform work, as workers nostalgically describe it.[31] But those days did not last long. The city's streets were quickly flooded with bikers wearing Caviar-branded orange thermal backpacks, all participating in what, in retrospect, seem less like "golden days" than a veritable gold rush—a boomtown of on-demand services. And it was not just Caviar. Zoomer, Grubhub, Seamless, Postmates, Instacart, the locally owned GoPuff—a whole spate of start-ups seemed to launch in Philly's now-bustling delivery market around the same time, and couriers seemed to be everywhere you looked.

Within just a few months on the job, I had already become familiar with the contradictions that Nunberg associated with gig work. I learned that you could log onto the platform whenever you wanted, but that logging on did not guarantee that you would make any money. You could reject any order you did not want to take, but those rejections seemed to ensure that you would not get another job for a while (sometimes not for a full hour). What to do during that downtime? Some couriers liked to hang out in the tony Rittenhouse Square while they waited for their next order. I wound up spending a lot of money on coffee, purchasing the opportunity to duck inside a café, out of the cold or heat. If I am being honest, I probably would have stopped delivering for Caviar after only a couple months had I not decided to study the platform. Most couriers I met were in college or had other jobs. But I also met workers who had decided to make delivering for Caviar their primary source of income. For these full-timers, the contradictions of gigging were becoming an everyday reality.

Nunberg's words did not resonate with everyone, however. Among champions of the platform economy, caution about the rise of gig work was nothing more than "blinkered nostalgia" for the pitiless nine-to-fives that Americans had suffered for generations. Writing for the conservative Manhattan Institute's economic policy blog, Robert Graboyes argued that the flexibility of platform work presented a liberating alternative to the repressive conditions that Marx and Engels observed on the factory floors of Manchester, only now manifest in the cubicles of office buildings and industrial parks. "Anti-giggers," Graboyes lamented, were "romanticizing and reinforcing the most hated aspects of post-1800 employment—subordination of and control over employees by employers." The security and benefits that traditional employment afforded simply cost too much in terms of autonomy. Platforms, by contrast, promised "the flexibility to continuously upgrade skills in a rapidly changing economy or pursue dreams that are incompatible with a 9-to-5 routine."[32]

Of course, flexibility is attractive to workers for good reason. Being "your own boss," you do not need to ask permission to take time off, and this means the freedom to invest in more meaningful and worthwhile activities—"to attend classes, acquire new skills, compose music, or invent the next killer app or cancer cure," as Graboyes suggests.[33] All those activities certainly sound more appealing than work. But curiously absent from the arguments was the fact that "upskilling" or curing cancer requires having the resources to support yourself—and that means getting paid, which was not a guarantee on the platform.

If anti-giggers romanticized the traditional employment model, then platforms tapped into an equally romantic sense of freedom as they recruited workers in urban markets across the United States. Uber blasted Facebook and Craigslist with ads promising drivers the freedom to "work when you want" and "choose your own hours." Caviar used a similar strategy in its recruitment materials. A video posted to Caviar's onboarding page shows a bike courier on a pleasant set of deliveries through San Francisco, dropping off food and smiling at customers, as he narrates his appreciation for the work's flexibility:

> I love discovering new places to eat around the city and sharing the best local food with people who appreciate it. I also love being in control of my schedule. As a Caviar courier, I work as much as I like, which helps a lot. I can easily balance work and school, and I can hit the books when I need to. Once I'm available again, orders are dispatched to my smartphone. All I have to do is accept.[34]

If all this flexibility sounds too good to be true, it is for good reason. My brother and I commiserated regularly about our experiences. We encountered long wait times at restaurants, hours spent idle between orders while not getting paid, and confounding policies for emergency situations

Figure 12. A still from one of Caviar's worker recruitment ads depicts a bike courier wearing the company's branded orange thermal backpacks. The ad invites viewers to "see what couriers say about Caviar," with narrators celebrating the job's flexibility. Because couriers are independent contractors, Caviar cannot require its workers to wear the bag. When I started work as a courier, the company gave the bags out for free as a way to increase the brand's visibility. The firm now implements insulation requirements but makes workers purchase their own bags. Branded carriers are now for sale. Video by Caviar (https://vimeo.com/117130396).

like getting a flat tire or finding a restaurant unexpectedly closed. The city became an obstacle course, and we studied its ins and outs. I can now read an address and tell you which side of the street a building is on (odd numbers on the north and east sides, even numbers on the south and west). I got to know Philly's dumpster-lined back alleys, which served as useful shortcuts to avoid rush-hour traffic. I cataloged restaurants that made me wait outside in the freezing cold, that would not let me use the restroom, or that took more time than allotted to get an order ready. I instinctively memorized the addresses of the apartment complexes and condominiums with the slowest security procedures—all barriers to making the delivery and getting another order.

With this local knowledge, my brother and I came to understand that the rhetoric of "flexibility" shrouded as much as it revealed about the realities of platform work. Flexibility taps into the cultural appeal of freelancing while glossing over the precarity of not having a contract or guaranteed pay: on nearly every on-demand service platform, independent-contractor workers are hired on a "pay-as-you-go" and "zero-contract" basis.[35] You earn money only when something needs doing—a piece rate for discrete tasks. Learning the shortcuts can help you get around more quickly, but it cannot guarantee the next order. Anyone can enjoy flexibility's benefits—all you have to do is accept these terms and conditions, which always apply.

Labor scholar Ursula Huws theorizes flexibility as a cornerstone of a new paradigm in the moral economy and organization of work—what she calls "logged labor."[36] Logged labor ties income to the demands of constant availability and monitoring; it pegs flexibility's benefits against workers' security. According to Huws, platform work is "logged" in three distinct but related senses: it is standardized into discrete components (a log of tasks), subjected to ongoing surveillance (a log of worker activity), and accessible only through online platforms that require constant connectivity (one must be logged on). Individually, these arrangements are hardly novel—as one colorful commenter put it, platform work is just "the same old piss in a different bottle."[37] What it describes, rather, is the convergence and intensification of these trends at a specific historical juncture—the fallout from the 2007–8 financial crisis.

Despite fiscal policies designed to encourage spending amid the economic downturn, businesses and public agencies felt the need to avoid large expenditures, including even minor investments in "human capital." Unemployment soared as a result. But even in the throes of the Great Recession, work still needed to get done. The tech sector stepped up to the challenge, with companies like IBM transitioning from hardware production to enterprise software; meanwhile, consultants like Deloitte and McKinsey began advising employers on how to use technology to amplify worker productivity and "streamline" the workforce.[38] With a host of new tools coming to market, firms found that they could essentially have their cake and eat it, too: management platforms and other online tools would give management the "flexibility" to avoid long-term employment commitments while keeping labor abundantly available, "ready to respond whenever their services are required"[39]—now at a piece rate and as a third-party service.[40] We see this impulse explicitly in a particularly chilling description of crowd-working platform CrowdFlower, given by its CEO:

> Before the internet, it would be really difficult to find someone, sit them down for ten minutes and get them to work for you, and then fire them after those ten minutes. But with technology, you can actually find them, pay them the tiny amount of money, and then get rid of them when you don't need them anymore.[41]

Although recruitment materials would never concede it, the same logics dictate the organization of work on service platforms—from restaurant delivery and grocery shopping to ride-hailing. Uber's Travis Kalanick, for instance, once described his firm's business model as making "transportation as reliable as running water": like the tap, workers are available when you need them and cost nothing when not.[42]

According to Huws, one of the most direct influences on logged labor are the "just-in-time" methods for staffing and operations first pioneered in the Japanese automotive industry and later extended to the factory floors and distribution centers of the world's vast supply chain networks.[43] Just-in-time reconfigures the organizational logics of work to accommodate a more dynamic management style, with labor and materials ordered up to meet demand, but nothing more. "Faced with fluctuating demand for labour," Huws writes, employers at warehouses, supermarkets, and delivery, security, and cleaning services "have for several decades resorted to just-in-time methods to allocate staff to shifts or add capacity at short notice." This allows firms to stay "agile" and flexible enough to gain a competitive advantage over rivals.[44] Streamlining is paramount, with workflows pared down to their most elemental components—a sort of Taylorism-plus for the entire operation. According to political economist Andrew Sayer, just-in-time management is grounded a simple philosophy of "waste" elimination—wasted labor, wasted product, idle time, surplus workers, and so on.[45] "Instead of producing at maximum volume in long runs in anticipation of demand," as was the norm in assembly-line mass production, "the essence of [just-in-time production] is that work is only done when needed, in the necessary quantity at the necessary time."[46]

The parallels between the just-in-time methods of logged labor and what my brother and I experienced on the Caviar platform are hard to ignore, even if the lexicon has moved on to "gigging." It is no coincidence that critical logistics scholar Deborah Cowen describes work along global supply chains as "just-in-time jobs" while legal scholar Valerio De Stefano refers to platform workers as a "just-in-time workforce."[47] In both cases, the managerial logics of flexibility penetrate work's organization, undermining the stability and security of the Fordist economy's social contract.

How did we get from just-in-time manufacturing to ride-hailing and food delivery? The shortest, straightest path is cut by the discursive appeal of "flexibility." Flexibility is rhetorically lithe enough to accommodate nearly any appropriation of its positive connotations. For instance, while some use flexibility to critique increasingly vulnerable and precarious working conditions (especially for women and people of color), others celebrate "structural flexibility" as a strategy of economic resilience in the face of uncertainty.[48] Following Uber's lead, on-demand startups capitalized on this ambiguity to recruit their workforces, exploiting the surge in unemployment following the 2008 financial crisis.[49] As the Caviar commercial above illustrates, firms understood that scheduling flexibility in particular would be a key motivator for potential recruits, especially with a majority of early platform workers coming from the ranks of the shrinking middle classes.[50] A 2015 in-house survey of Uber drivers, for example, found that most worked for supplemental income or used the "gig" as a bridge between longer-term employment. About 30 percent of Uber drivers retained full-time jobs, while another 30 percent were employed part-time—all used Uber to supplement their income.[51] A 2019 study by researchers at the IRS yielded similar results, showing that platform work was secondary work for the majority of 1099-filers.[52] With real wages stagnant and inequality soaring, flexibility did not appeal to workers because they want to "upskill," as Graboyes and other platform advocates argued. Rather, it was a necessary condition for an already vulnerable workforce that needed to supplement otherwise insufficient incomes. Per the IRS study, the only observable growth in gig work (based on 1099 filings) since 2007 has been among those workers who took up a "side hustle" to bolster income from traditional W-2 work.[53] Flexibility is thus a prerequisite for many platform workers, not some added bonus. As the prospects of a decent, long-term, and secure job dwindle in the economy at large, platforms surfaced to sop up growing demand for additional revenue sources, dangling the promise of autonomy before those facing economic uncertainty.

On-demand firms go to great lengths to set the terms of "flexibility," intimating that it is the workers who benefit most from the platforms' generous accommodations. But workers learn quickly that it is the platforms themselves that reap the lion's share of flexibility's benefits. With an independent contractor workforce, the platform has the option *not* to pay workers between tasks—the service might be "on-demand" for customers, but it's "just-in-time" for workers. Equally important, though, platforms *outsource*—a lot—and this gives them an organizational advantage over traditional rivals in the service industry. Platforms subcontract nearly every function needed to fulfill a service, from the labor to the materials and infrastructures necessary to move people or goods from point A to point B. And with all this flexibility built into the business model, platforms do not really *produce* anything of value. As Ugo Rossi argues, platforms have merely perfected the art of *capturing* and *seizing* value that is "co-created through the productive interaction of digital technology, human labour, and the social ecologies of the metropolis."[54] Geographer Lizzie Richardson observed as much in her empirical study of the "flexible arrangement" of on-demand food delivery. The platform (in this case, Deliveroo) stakes its claims on having coordinated transactions between consumers and independent service providers. But this coordination depends on a strategically "flexible" negotiation of outsourcing (costs) and appropriation (value). Delivery platforms "decentralise their features through external networks—such as restaurants or delivery riders—and simultaneously recentralise content from these networks according to the platform's own formatting—such as online menu choices, or delivery slots."[55] This reformatting structures transactions to the firms' neo-Taylorist impulse, discretizing tasks to their most elemental components and extracting rent for the "value-added" service.

Unlike traditional just-in-time methods in manufacturing, then, the platform is not streamlining an existing workflow by eliminating waste; rather, it is creating new pathways for extraction from the transactions it facilitates and appropriates.[56] The strategic interplay between the outsourcing of costs and the appropriation of value plays out in nearly every facet of platform work, often with harsh consequences for workers. Uber and Lyft drivers supply their own cars of course, freeing the firms from having to invest in a fleet, but that has not stopped the companies from imposing requirements on the vehicles;[57] and because the firm is prohibited by regulations from exerting direct control over drivers' behaviors, they outsource oversight to customers through ratings and reviews (which can then be internalized as cause for discipline).[58] Delivery platform Postmates cannot punish a courier, say, for her bike chain being too rusty or loose, her clothes being too ragged or dirty, or for not wearing a helmet; but then again, the firm was never on the hook for the bike in the first place, nor does it supply uniforms or pay for health insurance.

In fact, only in the wake of major accidents, deaths, or other scandals have platforms agreed to forfeit any of their organizational flexibility and provide even a modicum of support for workers by taking on some of the risk. In 2013, an Uber driver struck and killed a pedestrian in San Francisco, raising questions about the firm's liability.[59] Although the driver was logged in and had the app turned on at the time, he was not carrying a passenger. Uber's legal team seized on this detail to shirk responsibility, claiming that the driver was not working at the precise moment the accident occurred. Only under scrutiny from lawmakers did the company concede and begin providing drivers full-time insurance. Lyft still does not fully cover accidents that occur between rides.[60] For its part, Caviar recently started covering courier injuries, but this was only after courier

Figure 13. Banner hanging along abandoned train tracks over the intersection where Caviar courier Pablo Avendano was struck and killed on the job. Protests after Avendano's death attracted national media attention, prompting Caviar to start offering its riders accident insurance for on-the-job injuries. The banner has since been removed but a "ghost bike" memorial remains at the site, with flowers and commemorative notes left regularly. Photograph courtesy of Mar Escalante.

Pablo Avendano was killed on the job on a rainy night in Philadelphia and ensuing protests attracted the attention of national media outlets, reflecting poorly on the company (see Figure 13).[61] Most platforms simply send condolences.[62]

The Vanishing Hand of the (On-Demand) Firm

How have platforms managed to hoard the majority of flexibility's benefits? At least part of the answer has to do with firms' self-styled image as technology producers, not service providers—a claim that advances the startups' anti-regulatory agenda. Postmates' terms of service, for example, states that the firm "provides a technology platform facilitating the transmission of orders by consumers to Merchants for pickup or delivery by Couriers."[63] What Postmates is *not*, the agreement reads, is "a retail store, restaurant, food delivery platform, merchandise delivery platform, or food preparation entity." If it *were* any of these things (which is indeed how Postmates is described in the press),[64] then the company would be on the hook for a whole range of worker protections and expenses. Ron Tal, a data engineer at Uber, makes a similar case for his employer:

> A taxi company contracts drivers, deals with vehicles, pre book rides [*sic*], etc. Uber deals with building data centers, running real time software services, facilitating payment and conducting research into the economics of real time transportation automation, among solving all sorts of other interesting technological problems—all things that are not done by a taxi service. It's a totally different operation from what a taxi company or a transportation service does . . . Uber is not a taxi company, but a *technology company that provides solutions for people's transportation needs*.[65]

Such boundary work is strategic, perhaps even cynical. But it certainly shores up the firms' tech credentials, which cannot be separated from their attempt to skirt regulations. And in large part the platforms have been successful in this mission: the courts have generally failed to rule in misclassification suits, leaving it to the firms to set the tone of their own regulatory fate.[66] As the logic goes, Uber and other service "innovate solutions"; they do not give rides or make deliveries. Therefore, why should they be responsible for independent drivers or couriers?

Like flexibility's rhetorical power to overflow any single frame of reference, the discursive gravity of "technology" and "innovation" exceed the regulatory and judicial battles into which they have been enrolled. In fact, the same justifications that Tal gives for Uber's exceptionalism come up consistently across *economic* accounts of the platform economy. It turns out, economists very much enjoy analyzing Uber and other platforms—and even this may be an understatement. As journalist Alison Griswold quips, "It's hard to explain just how much economists love Uber." Popular economist Stephen Levitt concurs, suggesting that Uber represents "the embodiment of what the economists would like the economy to look like."[67] Why are economists so taken with the platform economy? Why do platforms represent "the closest you can get to taking the pure economic theory of textbooks and summoning it to life"?[68]

While I cannot pretend to know why economists love the things they do, I can say that when they are asked to give an account of platform economics, economists tend to fall back on a particular variety of technological determinism—one in which, like Uber's self-image, platforms embody technical innovation and progress, not a new paradigm in work's organization. This detail is not to be overlooked. By causally linking the business of "disruption" to technology, economic explanations grant undue credibility to platforms' anti-regulatory agendas. Economists provide the intellectual justification for the Uber City. The argument is not merely that platforms should not be regulated (most generously, because "certain traditional regulations become redundant by platform innovations").[69] No, because economists valorize platforms as the technological distillation of broader and *normatively desirable* innovations in work's organization, they sanction the platforms' advance of market logics into the seemingly infinite horizons of the lifeworld. Platformization means marketization, and economists love markets. If we can now use our phones "to book a house in Timbuktu,"[70] then the notion that a platform like Uber could improve healthcare or education services does not seem so far-fetched.[71]

Economists' accounts of platform economics start from precepts first articulated in Ronald Coase's seminal theory of the firm. Writing in the 1930s, Coase observed a disturbing misalignment between economic theory's fundamentals and real-world market activity. "In economic theory," Coase wrote, "we find that the allocation of factors of production between different uses is determined by the *price mechanism*." But this was not what was happening on the ground in corporate operations. "Within a firm, this description does not fit at all. . . . If a workman moves from department Y to department X, he does not go because of a change in relative prices, but because he is ordered to do so."[72] If the market is the most efficient means of coordinating production as economic theory holds, then why do economies *need* corporations and all the control that they wield over subordinates? Why do firms ("islands of conscious power") exist at all—why not rely solely on the market (an "ocean of unconscious cooperation") to coordinate production?

Coase's answer was that the modern firm becomes rational under a particular set of conditions—when the internal hierarchy of the firm becomes less expensive than the "transaction costs" of the price mechanism to coordinate economic activity. The firm makes sense, in other words, when hierarchy and bureaucracy are cheaper than subcontracts.[73] Coase's argument received empirical backing in the 1970s with the publication of business historian Alfred Chandler's *The Visible Hand.* Chandler's rich account demonstrates how and why administrative planning succeeded market coordination for U.S. firms in the period from the 1880s to 1920s.[74] With the railroad and telegraph networks connecting production centers across the continental United States,[75] the volume and scale of economic activity, combined with the speed and ease of communication, made coordination by internal administration cheaper than contracting through the market. The corporation, it seems, was not at all adverse to centralized economic planning in the way economists were. The result was what we think of today as the corporate firm: a massive, vertically integrated business organization, handling everything from manufacturing to distribution and sales.

As much as economists may have resented the modern firm's "visible hand" of control, its hegemony went unchallenged for decades, both in political and intellectual arenas. By the 1980s and '90s, however, the theory seemed to be unraveling. We were now witnessing a reversal of administrative power—a "vanishing hand" and "de-verticalization" of the firm, with internal control giving way to the "leaner" and more "agile" market-based coordination of distributed production, third-party providers, and sprawling international supply chains.[76]

According to economist Richard Langlois, the firm's de-verticalization begins with the economic deregulation and technological advances of the 1970s. These developments created the conditions necessary for firms to streamline—that is, they lowered transaction costs to the point that the Coasian choice between internalizing or outsourcing coordination shifted decisively toward the latter. To advocates, the de-verticalized firm heralded a return of the economy to a more pristine visage of the market. "In many respects," Langlois reflects, economic production in the twenty-first century "looks more like that of the antebellum era than like that of the era of managerial capitalism. Production takes place in numerous distinct firms, whose outputs are coordinated through market exchange broadly understood."[77] In manufacturing, companies no longer had to make every part themselves; in the service industries, restaurants and hotels no longer had to do their own marketing or housekeeping—all non-core, non-branded functions were now to be allocated to whichever supplier or contractor offered the lowest bid.[78]

This is the very argument that the economists make for platform economics, only now told through the prism of the smartphone as the technology du jour. When Langlois was writing, it was the arrival of the dot-com internet and Web 2.0. For the platform economy, it is the mobile smartphone that lowers transaction costs.[79] From the pages of *The Economist*: "Now that most people carry computers in their pockets [smartphones] . . . the transaction costs involved in finding people to do things can be pushed a long way down. . . . Computer technology is producing an age of hyper-specialisation, as the process that Adam Smith observed in a pin factory in the 1760s is applied to more sophisticated jobs."[80] And with this regime of hyper-specialization, production requires new tools to coordinate production—whether that involves the manufacture of a commodity or the delivery of a service. As economic sociologist and consultant Esko Kilpi argues:

> The existence of high transaction costs outside firms led to the emergence of the firm as we know it, and management as we know it. . . . [But the] reverse side of Coase's argument is as important: If the (transaction) costs of exchanging value in the society at large go down drastically as is happening today, the form and logic of economic and organizational entities necessarily need to change. . . . Accordingly, a very different kind of management is needed when coordination can be performed without intermediaries with the help of new technologies. Apps can do now what managers used to do.[81]

New Orleans–based tech startup TrayAway, for example, uses mobile apps to manage third-party room-service and cleaning for major hotel conglomerates.[82] Like countless other "software-as-a-service" providers, the firm is staking its future on the hyper-specialization ushered in by technological advances. This is the smartphone as "vanishing hand," chipping away at the transaction costs that had made managerial planning and control the necessary evils of capitalism. Firms need no longer manage staffs; apps do that.

At least that is the economists' tale, spun out by tech firms and consultancies.[83] You will note in the quote from Kilpi that management does not disappear; it is simply displaced onto the app. This detail is left out of most "vanishing hand" accounts of the platform economy. In fact, as legal scholar Julia Tomassetti argues, although the narrative of the platform-as-disaggregated-firm "speaks in the argot of Coasian firm theory," it is "largely nonsensical within it."[84] On-demand firms did not give up control over their workers. They performed a bait and switch, invoking technology and innovation as a proxy for the "inscrutability of the market" while moving management online, onto our phones. Writing of Uber specifically, Tomassetti argues that a more accurate Coasian analysis shows that platforms do not "reduce the costs of market exchange between drivers and passengers," but rather "lower costs between Uber, as a seller of transportation services, and passengers." More precisely, the platform "reduce[s] *intra-firm* transaction costs to the point of making firm production cheaper than market production."[85] This is in fact the opposite of the vanishing hand argument. Technologies do not reduce the costs of exchange *between* firms; they reduce the costs of coordination *within* them. Nothing has "vanished." It is the "visible hand" in a different guise. Technology did not disaggregate the firm; it *internalized market coordination within the firm's operations*. Control did not evaporate; it just more closely resembles market logics of rational, uncoerced choice.

Further gaps in the logic reveal themselves if we consider the range of information and communications techniques now commonly used to track, monitor, and control workers across the de-verticalized firm's dispersed sites of manufacture, production, and distribution—none of which makes an appearance in vanishing hand narratives.[86] Critical organizational sociologists Graham Sewell and Barry Wilkinson, for example, observed in the 1980s that the just-in-time management of the de-verticalized firm "both create[d] and demand[ed] new systems of surveillance."[87] With "lean," "agile," and "flexible" firms subcontracting previously in-house functions, contractors at each point in the supply chain had to find new ways to ensure consistency and quality in the workflow. All this additional coordination required more, not less, control. Manufacturers developed or commissioned new technologies suited to the task, including early uses of enterprise resource planning software, GPS telematics, and radio-frequency identification (RFiD) tags, all designed to render the physical spaces and bodies of production informatic, and thus subject to control.[88]

At the very moment when the Coasian firm seemed all but extinct, a more accurate account is that the firm was evolving—the "island of conscious power" becoming an archipelago of coordinative control. The technologies of just-in-time production are now indispensable tools for some of the world's largest companies—Amazon and Walmart, to name a couple.[89] The same surveillance and discipline that Sewell and Wilkinson observed in the 1980s continue to define the logistical work of "integrated supply chains." We see this, for instance, in political geographer Anja Kanngieser's analysis of "logistical governance," with RFiDs and GPS telematics digitally "tracking and tracing" warehouse pickers' physical movements to ensure efficiency, instill panoptic discipline, and dictate bodily movements according to the most efficient routes and routines.[90] Fully integrated into production and distribution processes, these same technologies allow "downstream" firms to monitor and intervene in "upstream" circulations and flows.[91] Walmart was an early proponent of the universal product code (or, more commonly, barcode) and wielded its market power to require all its suppliers to install UPCs on their inventory. This granted the "superstore" access to massive troves of data on the movements of products, and therefore also on workers and customers—from the production line all the way to the point of sale.[92] Walmart may appear to be de-verticalized (insofar as it outsources the production of its goods), but it wields an immense amount of control over its nominally "independent" suppliers.

Far from eliminating managerial power, then, the technologies of corporate de-verticalization enhanced control, rendered it more efficient, more granular. They did not eliminate centralized management; they ushered in a more pervasive and subtler formatting of management at every point of production and distribution: the factory floor, the wharf, the shipping routes, the distribution center, and now the logistics of the "last mile"—on-demand ride-hailing and delivery. The smartphone comes ready equipped with many of the same tracking and tracing functions that Kanngieser observed in warehouse picking. Even as consumer devices, smartphones track and trace users' activity, registering movement and relaying that information to distant databases, augmenting interaction with the environment.[93] As Ned Rossiter argues, locative media *are* logistical media.[94] With smartphones, what was once the informatic purview of the factory and warehouse has exploded onto the city with the normative force of economics behind it. The platform economy takes up the mantle of the de-verticalized firm, not only by integrating technology into the workflow but by propagating the fiction of de-verticalization.

The Visible Hand of On-Demand Work

The "vanishing hand," then, is a useful illusion. Planning and control may be less "visible" on the platform than in the traditional firm, but they are far from gone. The same logistical techniques that Kanngieser observed in the warehouse, and that Sewell and Wilkinson saw on the factory floor, have simply been ported onto workers' mobile apps, creating a warehouse-like interface of informatic and physical space across the market territory of the city.

In what follows, we hear from platform workers about their experiences navigating this interface. After over a year and a half participating in and observing on-demand labor as a courier, I conducted semi-structured interviews with two dozen platform workers—eighteen couriers on Caviar and Postmates, and six drivers with Uber and Lyft.[95] At a moment when the norms and obligations of logged labor were becoming the "new normal," workers experienced "flexibility" as a moral benchmark. Lacking fundamental protections as independent contractors, platform workers were being asked to accept more of the bitter with the sweet—to take on the risk and liability formerly shouldered by the employer in exchange for greater freedom and autonomy. But as we will see, the rhetoric did not match the experience. Platform work both creates and demands new forms and formats of control,[96] and workers register these controls as a breach in the gig's social contract.

The Changing Interface

Across the interviews, the concern that came up most frequently—and which aroused the greatest frustration—had to do with information flow. Because the worker-facing app is the primary point of contact between workers and platform, its design can affect the day-to-day experience of platform work. Updates to the app occurred regularly. In some cases, there might not be any observable change to the layout, only a subtle reconfiguration of the scheduling system or routing features. In other cases, information that was once available in a particular spot might be removed or otherwise shifted around within the apps' stepwise procedures, often without explanation.

At Caviar, a major change took place in September 2015 when the firm launched a new version of its worker-facing app, Courier Prime. Although Caviar had been beta testing the update on a voluntary basis, couriers were now forced to make the switch if they wished to keep working. The most obvious change involved the presentation of information at the start of a delivery. On the previous version, workers could see the address of each stop along a delivery route when first prompted to accept an order. After the update, only the pickup location was available, with the destination hidden until the courier indicated that she was at the restaurant, that the order was ready, and that she was out for delivery. A company email explained the changes as a way to "simplify the way you view orders—making it easier to focus on the individual steps of the delivery and not worry about the rest."[97] But many couriers, it turned out, liked "worrying about the rest." They liked plotting their routes while riding to the restaurant or waiting for a food order to be prepared—especially if they had multiple, "stacked" orders in their queue, with two or three drop-offs from one or more restaurants. Mitchell,[98] who ran deliveries for both Caviar and Instacart, explained that removing the information made it difficult to "figure out timing":

> Like, especially if you already had an order in the works and now it's kind of like, can I even take [a new order]? Not knowing where it's going makes that just, like, more complicated. I mean, I guess I could take it. I can take anything. But sometimes it's not quite cost-effective for me—I guess, like, not knowing the time it's going to take. If it's effective, it makes more sense, but it's harder to know.[99]

Where the details of route planning had previously been the couriers' prerogative, the new app automated the process, and workers felt like they were losing control. The stepwise functions of the process were the same—couriers always had to "check in" at the restaurant and notify managers and customers that the order was out for delivery. But Courier Prime reconfigured these steps into a "need-to-know" arrangement, with certain details available only when the platform determined that you needed to know them.

This seemingly small change had larger ramifications. First, Philadelphia is a city of one-way streets, only a select few of which have bike lanes—and of those, even fewer are free from potholes. At the start of an order, the updated Courier Prime gave couriers a general idea of where a delivery was going by specifying the neighborhood, but not the specifics of the "last mile."[100] These details are important: one wrong turn down a one-way street can take you out of the way of your destination by several blocks. And if workers feel that they are losing time, it is not uncommon to catch them riding on the sidewalk or the wrong way down the street, putting themselves and others in danger. Second, because Courier Prime now withheld the drop-off location until the last minute, couriers lost the ability to screen for undesirable destinations. Many believed this was the true reason for the redesign. Couriers instinctively remember the street addresses of apartment buildings or high-rise offices if they encounter a problem there, especially when security guards are rude or if the place requires lengthy procedures to get to the customer. Some buildings, for instance, require visitors to wear a temporary identification badge that must be printed before going up to the customer's apartment; some make couriers take a service elevator only accessible from a separate entrance, which might be around the corner. Such procedures become impediments to speedy turnarounds, and speed is crucial if you want to earn a decent hourly rate.

Courier Prime's information overhaul came up repeatedly in my interviews, both among workers with little-to-no experience as bike messengers and with the more seasoned cyclists. Mitchell, who only started running deliveries when the platforms came to town, believed that the changes were meant to stop couriers from rejecting orders to "annoying buildings," "because it was a lot easier to deny something based on that information" on the old system. Mike, a dreadlocked cyclist with years of experience as a courier and bike messenger, also disapproved:

> I really liked it more before where you had all the information in front of you at all times. You knew when you were picking up, the exact address—where it was going to . . . Every change was "Hey, where's that information? Where's that information? Hey, I didn't verify my items yet so I don't know the exact address"—that's a problem. If I'm waiting on something, I want to plot my route. I don't want to have to, like, wait until I'm on my way to know where I'm going.[101]

Gary, who had prior experience delivering food for restaurants but not with an app-based delivery system, echoed the sentiment:

> When they switched around the app a couple months ago, it became a lot more opaque. They give the courier less information up front. We used to see the customer's address right away. Now we can't see the customer's address until we've checked in at the restaurant and clicked the button saying that we have all the food and that we're out for delivery. After that, it shows the address, which I find very frustrating because I used to spend my time waiting for the restaurants to prepare the food—I would plan my route. And I can't plan my route while I'm waiting anymore.[102]

Postmates workers also encountered changes to the app interface, in this case with an even more direct impact on earnings. Around the same time that Caviar switched to Courier Prime, Postmates updated both its worker- and customer-facing apps in a way that altered the tipping interaction. Suzanne, who worked for Postmates before switching to Caviar, explained that on Postmates, couriers make most of their money through tips—"because the payouts themselves are extremely low, like, the most you'll see . . . might be seven bucks, but it's almost never that, usually three to four dollars."[103] With the new tipping procedure, couriers noticed a decline in earnings immediately. That was when Suzanne decided to leave Postmates and instead deliver for Caviar. Karl explained the change:

> When I started at Postmates, there was a system where the customer would tip you like on the percentage, based on how much they paid. But they did away with that for just like a vague—like, the customer can just tip whatever. When I started, we'd have to pass the customer our phone and be like,

> "Could you sign here and like make sure it's all okay?" And they would do the tip then, like on how much they were paying, a percentage. But they changed that. Now they do it on *their* phones, kind of like Uber. And they don't get the same screen, it just lets them tip whatever, not like, "Click here for 20 percent or here for 25 percent or anything."[104]

If "apps can do now what managers used to do," then changes to the app mean changes in management and procedure, often without explanation and never with accountability. For workers who depend on platform-mediated interactions and information, these changes accumulate, altering working conditions and affecting earnings. Even seemingly minor updates can have a major impact on workflow.

Distance-Based Pricing Models

The couriers also observed updates to the payment algorithms, again, without warning. In addition to the interface redesign, Courier Prime also introduced distance-based pricing to Caviar deliveries. Although the manager at the Philadelphia Caviar office insisted that hourly earnings would not be affected, couriers noticed their income drop rather quickly. According to my calculations, my hourly average decreased from about twenty dollars to around twelve over the span of a few months. But beyond earnings, couriers also sensed that the new payment algorithm fundamentally altered the system's logic. Prior the update, workers earned ten dollars per order as a flat fee, regardless of the distance. And they liked this system—these were "the golden days." It did not matter if the delivery was near or far, workers could count on completing enough orders during the shift to make it worth their while.

Distance calculations rewrote the rules. On the new system, time was of the essence. Hourly earnings are constrained by a number of pragmatic limitations outside the worker's control—the time it takes a restaurant to prepare the food or to ride out to a delivery destination. These constraints sensitize workers to any factor that causes delays but is not accounted for in the payment. Traffic, for instance, can be an issue in high-density commercial areas like Philadelphia's Center City. With so many people concentrated near so many restaurants, a lot of orders come in for Center City. But because the distance between pick-up and drop-off locations can be extremely short—sometimes only a couple blocks—most of the orders do not pay much, around three to five dollars. With traffic in full effect, even those few blocks can be slow to navigate, and hourly earnings can drop to nine or ten dollars, which certainly does not feel "worth it" for those used to earning twenty to thirty dollars for the same amount of work.

Couriers also experience delays getting from the ground address to the customer, especially in high-rise apartment or office buildings. Deliveries to high-rises entail a time-consuming, ritualized sequence that even the speediest bikers begrudge—locking the bike outside, checking in with a security guard who may or may not have to look up the customer's phone number in a directory and call ahead, waiting for and then taking an elevator up, and so on. These small tasks accumulate throughout the day as time spent unpaid. In fact, it is not uncommon for bike couriers to post ironic selfies from elevators or building lobbies to social media with the tag "#messengerlife." Gary explained that in apartment buildings, any number of issues can slow you down, whether it is a slow elevator or incoherent security policies in the lobby. For example, while most security desks accept packages or mail for residents, they will not accept food deliveries on residents' behalf. From my field notes: *Delivery to apartment building. Called the customer, no answer. Security guard called up, no answer. I asked to leave the delivery with him; he said it's against policy. Took twelve minutes to finally get the customer on the phone. She then called security and asked for me to leave it there. Guard agreed.*

Couriers experience these and other policies as roadblocks before the next order. Suzanne has given the issue some thought and even raised it as a concern with the local manager:

> I have brought up [the high-rise] issue many times with [the Philadelphia courier manager].... I'm just like, "Yeah, okay, lunch time, yes, all of these orders are in very close proximity and that's why the payouts are low." But the thing is, it will actually take me *more time* to deliver from [a popular, vegetarian fast-food takeout restaurant] to a high-rise in Center City than riding all the way to South Philly and dropping it off at someone's doorstep.... Like, first, I have to go to reception, then I have to wait for them to call, then I have to go all the way up, then I have to come all the way down.

> I have to lock my bike and unlock my bike. Whereas, if I'm dropping off at a house in South Philly or West Philly [both residential neighborhoods about two miles away from Center City], you don't have to lock your bike. You just go to the door.... In the end, I much prefer spending that time riding my bike than an elevator.

It was not that workers felt they should not *have* to deliver to high-rises—as Suzanne put it, "It's just one of the struggles you have to deal with because there are long orders that I get that are better and take me less time." The problem was the algorithm's inability to account for these roadblocks in its payment calculations. And with the drop-off details now hidden until an order is out for delivery, workers felt like they had no control over the situation.

Unsurprisingly, many developed tactics to speed up slow encounters—tricks to keep things moving. Kevin, for example, developed a certain swagger that he finds helpful in getting to the customer faster:

> As often as I can, I try to just, you know, walk behind the doorman. Go right past. Just, like, walk with purpose and, I don't know, confidence, and maybe—I'd say pretty often—I'll just walk right by them. And it saves me time.... The one suggestion I have is to avoid the whole doorman thing if you can.... Ultimately, if it were up to me, I wouldn't go to apartment buildings because they take time and time is running, but you end up getting those orders.[105]

As Kevin intimates, the issue was not only the high-rise's verticality. It was the social distances and topologies that the high-rise manifests. As platforms hit the Philadelphia market, the city was in the midst of the largest development boom it has ever seen.[106] With young professionals and retired empty-nesters trading in their suburban townhouses for downtown condos, wealth was moving into what had previously been nonresidential commercial districts or lower-income neighborhoods. Newly constructed apartments seemed to pop up on every block.[107] Caviar's specialty of high-end restaurant delivery certainly catered to the preferences of the "new Philadelphia," so it was not surprising that many of the platform's customers lived in new high-end high-rises. But with this niche came competing objectives. While the courier, for her part, serves as a conduit of luxury—a cog in the infrastructure of convenience that the "on-demand" economy sells—the wealthy new residents also desired security. Like gated communities built into the city skyline, the new buildings' lobby procedures were designed to filter out those who belong from those who do not, setting up a confrontation between the infrastructure of convenience and the fortifications of luxury—all barriers to speedy turnover.

And the new high-rises were not the only places where Philly's changing demographics created friction. Outside of Center City, couriers also ran deliveries to quickly gentrifying low-rise residential neighborhoods. As Suzanne indicated, couriers generally appreciate these orders since they mean more time spent riding a bike than an elevator. The problem, however, is that most of the restaurants in Caviar's network were concentrated in Center City. On the new distance-based pay system, it became all but impossible to get a follow-up order if you were delivering outside the downtown zone to the burgeoning enclaves of new wealth. Couriers thus found themselves having to return to Center City, unpaid, before getting their next order. Mike called these "one-way jobs"—long runs where you are unlikely to get a follow-up. "It's like you're doing all those trips and then coming back down and not getting paid and then getting sent back out to do the same thing again."

The same issue can also come up at the start of a shift. Prior to the Prime update, couriers could log on at home and receive their first order within a reasonable amount of time. This changed with distance-based pay, likely because the algorithm was designed to allocate orders to the courier closest to the pick-up. If you did not live in or near Center City, you might have to travel before getting an order. As Suzanne explains:

> Since I live in West Philly, a lot of times I'll just sign on at my house and wait. But I've gotten text messages saying, "Hey, can you head towards Center City?" And I'll just be, like, come on! . . . I know you can send me an order for Center City while I'm out here, you've done it before. It's just, you're gonna make me ride into the city so you can give me a lower-paying order. I mean, when they changed the payout system to be based on location and distance . . . I could see how people could really try to abuse that

> system in the beginning. But I don't think they can anymore, really. The downside is that [orders] are now only going to go to couriers already in the area.

Whether it is high-rises, one-way jobs, or a queuing system that privileged certain zones, distance became for workers a crude and, as we will see in the next section, increasingly incoherent measure of labor. As Suzanne quipped, "It's cheaper to do that, so they will."

The Right to Reject

As independent contractors, platform workers reserve the right to turn down any job order they wish. Some couriers chose to reject if the pick-up was in a particular part of the city. Marcus told me that he avoided delivering in Center City at night because he liked to "stay in the neighborhoods." Others expressed the opposite preference, preferring to avoid Center City during the day because the traffic makes it hard to get around. Most couriers, however, explained that they reject orders when the payouts "just don't make sense"—a common refrain. I asked Alexandra, a Caviar courier who delivers by car, not bike, if she ever rejected:

> I do occasionally, if they don't make sense. Sometimes I'll get really ridiculously low-paying orders that are actually really far, or it's a restaurant that I know takes forever and that [the price] doesn't make any sense. Usually, I'll text dispatch and be like, "That payout doesn't make sense."

Charlotte and Matt, both seasoned cyclists before starting at Caviar, expressed a similar sentiment. I asked Charlotte if she felt like she had a sense of how the algorithm assigns and prices the orders:

> They go over it with you somewhat, but I feel like it's still one of those things where you're not getting this breakdown every time you take a delivery . . . like, there's no itemized receipt or anything. Some of them are pretty far, but the money you're getting is pretty low, so it doesn't make sense. Because for others, you're not going as far but you're making more money and it's kind of, like, what? I feel like they need to make it—like, if you're taking a small order but going a longer distance, you still should

> get a good amount of money, whereas if you're taking a bigger order, even if it's a shorter distance, sometimes you get paid more and that doesn't make any sense. . . . The farther you go, the more risk you're putting yourself in. If you're in a blizzard and you're going from Center City to West Philly, you should be making good money on that because that can be dangerous. . . . So sometimes those payouts just don't make sense in terms of safety.[108]

Charlotte was especially attuned to risk and safety. When I interviewed her at a coffee shop near her home, her foot was in a cast, propped up on the chair next to her. A few weeks earlier, Caviar halted operations due to a snowstorm; when the platform reopened, Charlotte logged on to catch up on earnings. Although the snow was mostly cleared from the streets, a combination of black ice, exposed trolley tracks, and reckless driving resulted in a fall that left Charlotte with a broken ankle. Not only was she now a couple thousand dollars in debt due to deductibles on the insurance that she pays out of pocket, she was also no longer able to earn money because she could not ride. When we finished the interview, she told me she was staying at the café to look for a new job on her laptop.

Matt, who had years of experience as a bike messenger in both Philadelphia and New York City, also explained that he rejects orders that "don't make sense." "Weird" payouts and slow restaurants were all candidates. But he also pointed out that there could be consequences for rejecting orders too often: "In general, I try to accept most orders because who knows if they're tracking your rejects."[109] A similar uncertainty came up for other couriers, sometimes manifest as a sense of anxiety around the platform's metrics. The managers offered little insight into the data the platform tracked, and specifically, whether your "reject rate" affected your status in the queue. Mike, who, like Matt, had years of experience as a bike messenger, felt that rejecting too often could trigger something within the system to automatically set you back in the job order: "I'm almost sure that there's a rule, if someone denies more than one job in an hour, ya know? . . . There's definitely a queue at all messenger services and you go to the bottom. Same at Caviar. I'm positive." Less seasoned riders, like Janet, sensed the opposite—that rejecting orders could somehow "communicate" or signal your preferences to the dispatching algorithm:

"If enough people reject the really bad orders then they'll stop doing them. I've tried to do that and sometimes it works."[110]

Whether it is paranoia that the platform is tracking your rejections or a hopeful sense that it might tailor to your preferences, both responses are guesses at best. The platform managers refused to disclose the technical aspects of queuing and rejections, likely because they hoped to prevent workers from "gaming" the system. Ride-hailing platforms Uber and Lyft are curiously more candid about rejection policies. Both platforms require drivers to maintain a certain acceptance rate; falling below the threshold becomes grounds for "deactivation"—getting locked out of the platform, sometimes for a couple hours, sometimes for weeks or even permanently.

But while ride-hailing drivers did not experience uncertainty in this regard, the platforms' acceptance-rate policies created other anxieties. According to Dante, a freelance web developer who had been driving for both Lyft and Uber for about a year, "They're straightforward with you, the platforms. Like, they'll send you notifications saying you're dropping below your threshold or whatever, but sometimes they also will just change up the policy." Lyft, Dante explained, allots a certain number of "directional" jobs each day, with drivers indicating a preference to take a ride in the direction of their home at the end of a shift. Without explanation or notice, Dante witnessed the allotment drop from six to two per day. While the change did not affect him much, Dante found it to be indicative of the platform's power to "change the rules."

Maria, a forty-year-old mother of four who had also been driving for Lyft for about a year, felt the anxiety of maintaining the acceptance rate on a bodily level. She told me that she often logs off immediately after dropping off a passenger. If you forget to log off and then miss a ride, she explained, you will be penalized. She even avoids drinking too much water or Gatorade while driving, "because if I have to pee, I have to log off. You can't be signed in and get an order and not take it. They count that against you. If you need gas, same thing—you don't want to have to stop when somebody's waiting for you to pick them up. So, like, after I drop someone off, I'm usually logging out for a minute, just to take a quick break, like, get off the road, because sometimes I've been on for hours."[111]

Ensuring Coverage

Despite promises of scheduling flexibility, only certain times of the day or days of the week are profitable—weekend evenings are generally busy and weekday mornings are generally slow for food delivery, for example. Workers learn these dynamics quickly. But beyond such naturally occurring variability in demand, workers' scheduling can be dictated by "nudges" and incentives that the platform uses to ensure coverage.[112] When Caviar rolled out the Courier Prime update, the app included a new scheduling system. Workers were now being asked to sign up ahead of time to "get priority," or preference in the queue. The scheduling interface shows a week ahead, with each day broken up into hour-long shifts. The busiest shifts display "peak pay" bonuses, highlighted in green, sometimes up to 20 or 25 percent above base pay for the busiest times of day. Once a slot hits its quota, the shift is crossed off and marked as full.

As Mitchell lamented, incentives like peak pay have to be understood in relation to constantly declining base rates. The bonuses help, Mitchell explained, but it would be better to just raise the baseline:

> I feel like the only thing they've told us about is when it's pay *increases*, either on bonuses or at peak times. I know they added in peak times to it, at least, supposedly, and it seems like they are in effect because I have done afternoon deliveries and [the payouts] seem super fucking low compared to when I'm doing dinner rush stuff. At dinner, it's usually seven to fifteen bucks, whereas during, like, an afternoon shift, I only get half the orders and we're going to get max six bucks, even if it's far.

By its nature, the scheduling system attempts to impose some order on the chaos of a self-scheduling workforce. The system allows the platform to plan ahead, to anticipate fluctuations in supply and demand, incentivizing work during the busiest periods and disincentivizing slow times of the day. The trouble, as Mitchell observed, is that the disincentives are far less transparent than the incentives. The platform shows what the managers want workers to see.

Getting workers to stick to the schedule, however, requires some pressure. Caviar's terms of service state that workers can always log on, even if they are not prescheduled. Those who preschedule will receive preferential treatment—they will "get priority." Naturally, this frustrated workers like Paul, who had been riding for the platform since the more freewheeling "golden days." When I interviewed him, Paul was finishing up college and he complained that the scheduling system made it more difficult to balance work and other commitments:

> My main thing—I guess I have an issue with the "priority" working, like not getting "priority" if you just sign on. I understand why that makes sense, but like . . . I just know a lot of people who are in college working for Caviar and it's just not always possible for me to know a few days in advance if I'm going to be able to work, compared to—I just don't know what my workload is going to be like week to week, because I got a lot of professors who change up their schedules. The whole point was to be able to work whenever, right?[113]

Gary also observed the scheduling policies become more stringent. He pointed out a "reliability index" that Caviar started including in its weekly emails to couriers. The index visualizes the portion of time that workers spent logged on during their prescheduled shifts, essentially chiding workers for signing up but not showing up. This created what Gary saw as an undue sense of anxiety:

> When I signed up for Caviar, I was told that we weren't obligated to accept orders, that it's completely at our discretion when we want to work and what orders we want to accept. That was a big selling point for them looking for couriers: "Take this, do this flexible thing, you can just do it in your spare time." Now, they're doing this reliability system. . . . What it feels like is Caviar is trying to guilt-trip us for not showing up for our shifts—which are not obligatory—and whether or not we're being penalized for showing up for our shifts is kind of unclear. But it seems like, whether or not they're penalizing us, it feels like they're asking us to penalize ourselves.

While it may be unrealistic for any job to be as flexible as Gary and Paul had imagined, their experiences provide a sharp contrast to the courier narrating Caviar's recruitment video, who found it so easy to "balance work and school" with the job and "hit the books as much as I need to." If the conceit of platform work is a trade-off between flexibility and security, then platform workers were being asked to accept the terms of those conditions but without the benefits.

Another anxiety is that shifts become overstaffed. Not knowing how many other couriers are working, Matt felt like he could be logging into an unknowable situation, with too many workers "diluting" demand. It would be useful, he explained, to see where other workers were located, like "dots on a map." At least that way he could try to position himself more smartly and avoid the areas already crowded with Caviar workers. "I'm not going to go out just to go to where that herd is." When it is busy, Matt explained, "it doesn't matter—you're just getting order after order." But when it is slow, there are times when he wonders if he should even bother coming out: "Sometimes you go out, you ride by Rittenhouse and you see just how many [other couriers] are out there—like, I probably wouldn't have left the house to begin with if I knew that many were out there!"[114]

If too many workers on the road "dilutes" the demand for couriers, it is not an issue for the platform. With piece-rate wages, there is no cost to the firm for being overstaffed. An understaffed platform, by contrast, is an operational nightmare. Being understaffed, customers face long wait times, which can leave them dissatisfied, and it is all too easy for customers to switch to any one of Caviar's several rivals—Grubhub, DoorDash, Seamless, Postmates, and so on. With no costs to overstaffing, the platform can become trigger-happy and overcompensate for worker deficits. It is therefore not uncommon for Caviar dispatchers to send out notifications encouraging more workers to log on during demand spikes—what the company calls "the bat signal." Nudges like the bat signal coax workers to cash in on what dispatchers promise to be a busy night. But as we saw with Charlotte's injury, the same conditions that bump demand often put workers at risk. In fact, it was after a bat signal on a rainy night that courier Pablo Avendano was struck and killed on the job.[115]

According to journalist Juliana Feliciano Reyes, these nudges exemplify the growing sense of uneven obligations between platform and worker. When I responded to bat signals, it was common to receive one or two orders immediately but then hear nothing for an hour or more, sometimes for the rest of the night. Logging on in these situations, workers alleviate the staffing problem but then get left high and dry, sometimes miles from home and in the rain.

Bat signals can also be a problem for workers who prescheduled themselves to ensure "priority." For these workers, the bat signal draws out more workers than necessary and, as Mike puts it, the work just gets "scattered":

> If you send out a bat signal, that's fifty people that're going to sign on, like, within a few minutes, and then the work just gets scattered. That happened to me just the other day. It was like 10:50 a.m., a week or two ago; it's drizzling and I'm just putting my shoes on, getting ready to walk out the door for an 11:00 a.m. shift and I get a bat signal. I'm like, "Come on!" . . . I barely made anything . . . and I was *hustling,* didn't deny a single order all day. I texted the dispatcher, like "What's going on, where's the work at?" and they send me like a $3.50 job. So I was like, "It's $3.50. Come on! Let's do this, make me sweat."

Clearly, Mike was not only ready to work; he was ready to work hard. He followed the rules and signed up on the scheduling system ahead of time, only to sit around for much of his shift waiting for an order that paid well. He communicated these frustrations to the company through a feedback mechanism on the app:

> I put this in the little blurb and I was like, "I think when I started working here it felt like a bubble that I knew was going to burst and it did. It did, the bubble burst." And the response was very vague, like, "Oh, I hope the bubble hasn't burst. Sometimes it's just slow." . . . But how are you going to send the bat signal out right before a major shift starts? . . . Like, if it's slow, don't get fifty more couriers to sign on right before the big lunch shift starts! . . . The reason I schedule myself at Caviar the way I do is so that I know—I make pasta the night before and I'm making an omelet in the morning and I'm going to be riding all day. If I'm getting out of the shower and I get a bat signal, I'm not going to throw my clothes on and go out to make thirty dollars. Fuck that, I don't do that. So the bat signal for me is nothing more than "you're about to make 20 percent less money for the next couple of hours because everyone and their mom is signing on."

As Mike sensed, the platforms *were* in the squeeze of bubble-like economics.[116] Backed by venture-capital injections, on-demand firms entered markets like Philadelphia with unsustainable models, paying out high rates to build up their fleets and charging low prices to build up a customer base—prices that hardly reflected the true cost of a delivered meal or a ride home. Most platforms remain unprofitable, spending more to sustain their growth than they bring in.[117] The goal appears not to actually make money on the service, but to corner the market for it.[118] The trouble is, in the market for on-demand services, the competition is too strong and the profit margins too thin for the gambit to work.[119] Customers are not loyal—most deliveries can be fulfilled by a substitute platform—and neither are workers: many couriers and drivers work on multiple platforms at a time, hopping back and forth when it is slow or the pay is not good.[120] As Uber and Lyft disclosed ahead of their 2019 initial public offerings, the only viable options for platforms to become profitable are to either raise prices for customers or dramatically lower labor costs (or even get rid of workers altogether).[121] All evidence suggests that they are choosing the latter.[122]

The Rational (Flexible) Worker

To make it in the smart city, workers must navigate a complex entanglement of platform logics, financial gambits, and shifting social topologies. They develop intuitions, a sense of the city's rhythms and flows, but also a sense of the algorithm's calculations, and they must adapt to its rationality accordingly if they hope to earn a decent wage. But what does "rationality" mean when the market is itself irrational—when workers shoulder the bulk of the firm's overhead costs without guaranteed income? When labor is available "on demand" and "just in time," but with managerial policies that can change on a dime and without accountability?

Scholars of critical cultural studies have long asserted that economic rationality is a cultural artifact—that markets are thoroughly social and political entities.[123] Science and technology studies likewise insist that rationality is never the province of human reasoning alone—it emerges as the outcome of more or less stable configurations of institutions, scientific authority, and "market devices": "material arrangements, systems of measurement, and methods of displacement—or their absence."[124] In the smart city's economy, the platform sets those configurations. Workers may have the "opportunity" to determine for themselves whether a given task or order is worthwhile, but the firm dictates every input—how information is retrieved, how tasks are priced, how orders are allocated, and so on. The platform sets the parameters within which workers make decisions; it bends and contorts the options to the point that the only decent choice is that which serves the firm's interests. The platform, in short, uses market "rationality" as a managerial technique.[125] When workers act "rationally," they become calculable—their actions become legible and predictable, and thus manageable.[126]

At least as much can be inferred from platform workers' experiences. But by design, the platforms are opaque about their operations. On-demand firms claim trade secrecy protections to keep their data and price-setting formulae black-boxed and under wraps.[127] To prevent workers from "gaming" the system, platforms refuse to disclose queueing policies, or baseline pay rates, or the criteria for issuing "bat signals" and other nudges.[128] All this obfuscation makes it difficult to dissect and critique rationality's weaponization.

We encountered similar roadblocks in the previous chapter. With Intersection's executives unwilling to disclose how LinkNYC data support its advertising and marketing business, computer science modeling and patents offered insights. A similar tack can work for on-demand service platforms. In this case, however, it is the field of *management science,* which draws on economics, engineering, and computer science to provide a "scientific approach to solving management problems in order to help managers make better decisions."[129] Within management science, a growing body of literature tackles the challenges of labor management in the on-demand economy. This research constructs mathematical models and computer simulations of service platforms, mimicking the logistical challenges associated with "logged labor"—self-scheduling, customer impatience, matching challenges, and so on, all of which pose operational hurdles for the platform. Guaranteeing workers even the legally required minimum of independence makes service delivery logistically complicated, and the management science literature reveals how platforms can marshal workers' rationales against them to maximize profit.

For example, management researchers Gérard Cachon, Kaitlin Daniels, and Ruben Lobel argue that when workers can self-schedule, the platform faces two inefficiencies—understaffing and overstaffing (or what they call "demand rationing" and "capacity rationing," respectively). We saw these issues come up in the previous section. Being understaffed, the platform risks dissatisfied customers switching to a competitor; being overstaffed, the platform risks dissatisfied workers with long wait times between orders. To overcome these competing dilemmas, management science highlights the "control levers" that platform managers may "pull" to achieve the right balance between the two demands.[130] Cachon, Daniels, and Lobel identify the appropriate control lever in the payment contract, where "optimization" means tweaking inputs such that workers make the "right" decisions—assuring workers "sufficient expected profit" but without giving them "too much of an incentive to participate"; if the pay is too high, the platform faces "an excess of providers," but if it's too low, it "could entice too little participation from providers to satisfy demand."[131]

Striking the right balance is a difficult feat. Regardless of the exact calculations, the contract is deemed successful—efficient—only when it elicits an "elastic" response in the workforce, that is, if the incentive has the intended effect on workers' decisions to log on and thus the labor supply. In economics, labor supply is positively elastic if the wage adjustment is the same as the change in labor supply—a wage increase (decrease) that increases (decreases) the labor supply. By contrast, the workforce is negatively elastic if the effect goes in the opposite direction—a wage increase (decrease) that decreases (increases) labor supply.

Platform workers become "rational" when they exhibit positive elasticities. Positive elasticity means that workers are responding in *predictable* and *desirable* ways to the control lever as an incentive or disincentive. Workers that exhibit negative elasticities are deemed "irrational,"[132] and "irrational" workers do not make very much money. For instance, Uber drivers report that logging on when wage hikes—"surge pricing"—are not in effect leads to significantly lower earnings.[133] Many therefore find working outside of predictably busy periods to be "just not worth it." Harry, a driver for Uber and Lyft, explained his scheduling rationale:

> I've been driving pretty frequently. I would say two full evenings, afternoon and evenings. Like Friday and Saturday, I usually try to do from midafternoon all the way to three in the morning. . . . It's the only time that you can make really any decent money. . . . It's just not worth it to work during rush hour, even if there are surges, because there's just so many cars out there, it takes you forever to get anywhere. It's only really worth it if there are surges and no traffic, and that's really only when the bars let out.

If it's only "worth it" to log on when demand is highest (and, for Harry, when traffic is at its lowest), then it must also be the case that baseline, non-surge wages are functioning well as a *disincentive*. Cachon, Daniels, and Lobel discuss this explicitly in their optimal contract (i.e., not giving workers "too much of an incentive to participate"). But none of this was particularly obvious in the early days of on-demand services. Platform workers had to "learn" the right times and places to work if they wished to make decent money. A *Forbes* article from 2014 reports this learning in action for Uber drivers:

> For those working just a few hours here and there, earnings are wildly varied. Some drivers took home an impressive $60 an hour. Others earned less than $10 an hour. . . . Drivers are starting to figure this out. Expect to see more drivers working part-time on Uber, dropping in to work the highest-grossing hours but not wasting time on slow periods.[134]

In effect, both Harry and *Forbes* are describing the same thing—the "control lever" in action: the payment contract enticing workers to log on when and where demand is high and persuading them to log off when it is slow.

If platforms use workers' own "rationality" to satisfy efficient supply-demand matching in scheduling, they also discourage "irrational" behaviors. We see this in attempts to mitigate a practice known as "income targeting"—that is, when workers log off after hitting a predetermined earnings threshold or quota. To workers supplementing their income with a platform side gig, the practice makes complete sense—for a young father, it is hardly "irrational" to earn a target wage and then hurry home to get back to his family. To management scientists, however, wage quotas are negatively elastic, and therefore "irrational": as wages increase, workers hit their targets and choose to log off sooner, despite the prospect of higher earnings.[135] This threshold mentality threatens platforms' most basic assumptions about how workers make decisions. Michael Sheldon, data scientist for Uber, openly fears that practices like income targeting could "undermine the benefits of emerging 'sharing economy' markets where tasks are dynamically priced" and "significantly reduc[e] the economic gains" for the platform.[136] To discourage these behaviors, Uber barrages workers' phones with "nudges" and notifications when demand is high: "Are you sure you want to go offline? Demand is very high in your area. Make more money, don't stop now!"[137]

Similar issues come up when platforms modulate wages by geographic zone. Uber's "surge pricing," Lyft's "primetime," and Postmates' "blitz pricing" all follow the same logic of persuasion, grafting temporal staffing onto the spatial distribution of workers across the market territory. Surge systems subdivide urban markets into discrete zones and calculate supply–demand ratios locally; the platform then calibrates those ratios as price and wage modulations, broadcasting bonuses to workers on a map to incentivize movement into zones where demand is highest. Over time, workers learn from their experience to distrust this information. Seasoned Uber and Lyft drivers regularly advise novices in online forums that they should never "chase the surge."[138] Veterans know that trying to hit the hotspots can inadvertently "dilute" the supply–demand ratio and thereby "deflate" the wage adjustment (in the same way that the "bat signal" became a problem for Mike in the previous section). Charlie, a computer engineer and driver for Lyft, intuited as much. He even kept a running log to reverse engineer the platform's algorithm. His advice to new drivers? "Never go to a hotspot. Once you step in there, you kill it, you physically kill that thing, because you alter the ratio."[139]

Such advice is not likely to be well received by platforms, which depend on spatial incentives to coordinate their fleets. When workers reposition themselves according to the wage adjustments, they are demonstrating a sort of spatial elasticity. But if going to the hotspots has no benefit for workers, then chasing the surge clearly is not a rational choice in any meaningful sense, even if it obeys the "rationality" pushed by the platforms. Workers therefore develop their own calculus of the worthwhile, their own rationality, and this does not always line up with platforms' assumptions.

As workers' calculus evolves, management scientists are paying close attention. Operations researchers Harish Guda and Upender Subramanian, for example, have studied worker responses to platform incentives.[140] In a series of papers, including a publication in the field's leading journal *Management Science,* they challenge the industry norm of using surge zones to manage fleet distributions. Surges, Guda and Subramanian argue, are insufficient and potentially ineffective—not because workers are not responding to the surge, but because workers learn the demand patterns and preempt the surge, lingering in busy areas long after the demand has been satisfied. This creates new coordination problems of excessive supply (overstaffing) in certain areas and deficits (understaffing) in others. Instead of using surge pricing to *attract* drivers to market zone, they recommend that platforms "misreport" demand information to force movement *out of* saturated zones. Like internet service providers throttling bandwidth to police network abuse, this tactic—what Guda and Subramanian called "demand throttling"—raises prices to the point that customers stop requesting service and, with demand drying up, drivers leave the area of their own volition.

This cynical strategy depends entirely on workers' decision-making under information asymmetries. Because workers do not know where other workers are positioned in the market zone (they cannot see the "dots on a map," as Matt put it), they unknowingly impose what Guda and Subramanian call a "competitive externality" on one another. This makes it all but impossible for workers to position themselves strategically, and the result is exactly what drivers knew all along—that chasing the surge is not "rational." Note that Guda and Subramanian do not advise that platforms share more information with workers to help them avoid the competitive externality. Instead they propose "distort[ing] the price" in the oversaturated zones, "to deliberately choke demand" and "improve total platform profit across zones if the need for additional workers to move is sufficiently high."[141] In other words, by using workers' "irrationality" against them, "a surge price in the zone that the workers should *leave* [becomes] the less costly means for credible communication (than a surge price in the zone that workers should move to)."[142] Demand is satisfied, but the supply—the workers—are left with an entirely distorted view of the market that leaves little opportunity to make a decent hourly wage.

Logistics in the Urban Lifestyle Factory

When service platforms came to town in the early 2010s, the hype surrounding "gig work" was contagious. An "Uber for everything" craze ensued, and "there's an app for that" became a bad punchline. Washington, D.C., insiders even went so far as to propose a federal welfare overhaul based on Uber's logged labor model.[143] The platform economy represented innovation and entrepreneurialism, and these seemed like the cure for societal complacency. In Claudel and Ratti's terms, we were living in a Minitel world and platforms would "disrupt" it. Platforms seemed capable of translating *any* social problem into the language of the marketplace—and thereby opening social problems up to competition and progress. Indeed, markets are precisely what platforms are good at—a more efficient matching of supply and demand through the clever appropriation of computing power, networked communication, and otherwise-idle resources. To the extent that this combination of ingredients was indeed "innovative," it inspired a font of optimism among wonks and techies.

But when the recipe was operationalized, it was not at all innovative. The kind of "lifestyle" that Travis Kalanick associated with Uber—"give me what I want and give it to me right now"[144]—was about instant gratification, and anyone with experience in the service industry can tell you exactly how much labor is involved in the logistics of instant gratification. Though management now seemed to take place on an app, every managerial strategy had a precedent in the just-in-time methods of supply-chain manufacturing. Even the coordinative technology that platforms claimed to innovate was repackaged from the field of logistics.[145] The same could be said of platforms' business models. As Ugo Rossi reminds us, "the realisation of surplus value from labour through the reduction of wages, the application of new technological solutions and the intensification of the labour process"—in other words, the managerial strategies of platform economics—are also the "classic driver[s] of capital accumulation."[146] Zero-hour and "pay-as-you-go" contracts, risk-outsourcing, income instability, "sham self-employment," exclusion from workplace protections—

none of these trends is particular to the platform economy, yet all are characteristic of its logged labor business model.[147]

What, then, is so notably "innovative" about this "future of work"?

If anything is at all groundbreaking, it is the platforms' expansion of capitalism's extraction and seizure operations onto ever more granular and routine planes of life.[148] While capitalism has always forged new frontiers by enrolling previously nonmarket relations into flows of finance, labor, and commodity exchange, the logistics revolution amplified and legitimated these tendencies, "smash[ing] factory walls, moving productive and value-adding activities well beyond the delimited spaces of industrial activity."[149] The platform economy's "innovation" was to revive and repackage this logistical project in the guise of "tech," pushing capitalism's extractive operations onto the economic circuitry of everyday life, reformatting urban circulations and flows as infrastructures of convenience, instant gratification, and lifestyle.

We see the dynamic in microcosm in grocery-delivery startup Instacart's business model. Unlike its failed dot-com-era predecessor Webvan, which acquired warehouses to stock inventory, Instacart simply hires "shoppers" to purchase and deliver items from supermarkets to customers' doorsteps.[150] Sociologist Chelsea Wahl conducted ethnographic research as an in-store Instacart shopper, and her findings reveal remarkable similarities between working the aisles of a suburban supermarket fulfilling Instacart orders and the breakneck labor of "picking" at Amazon's fulfilment centers—perhaps the quintessential figure of contemporary logistical labor.[151] Just like the scanner devices that direct and track workers through Amazon's nearly million-square-foot warehouses, the Instacart shopper's smartphone app displays an anxiety-inducing countdown of the time allotted to locate each item on the customer's grocery list in the aisles of the supermarket.[152] The only difference is that Instacart's shoppers must learn and memorize the supermarket's layout, whereas the organization of Amazon's "optimized" inventory is not designed to be knowable or even legible to humans.

Like the platform economy as a whole, Instacart's "innovation" was to reformat social space. The platform co-opts an everyday context of consumption—the supermarket—and transforms it into a space of value capture and seizure. And if we allow the part to stand in for the whole, we begin to see how the platform economy reformats *urban* space. What Instacart does with the supermarket, so the platform economy does with the entire city. Writing in the 1960s, Manuel Castells argued that the modern city spatialized the "collective consumption" necessary for reproducing capitalism's labor power.[153] Progressive urban politics, consequently, had to grapple with decisions about how best to allocate resources for the collective good. The platform economy now repurposes the city to suit its own ends—the capture, seizure, and extraction of collectively produced economic value.[154] As a result, the politics changes. The city is no longer a space of collective consumption but a "context of control" to support not collective but *selective* consumption, the surfaces and circulations of everyday life now reformatted as an infrastructural support for Kalanick's "lifestyle" of instant gratification.[155] If consumption is itself increasingly a service, not a product or commodity, then service provision can no longer be untangled from the novel forms of extraction activated in its delivery. Under this regime of "logistical urbanism," as digital culture researcher Moritz Altenried calls it, production and consumption converge in the "last mile," where the "rationality of the integrated management of flows" spills into the city as the platform "reconfigur[es] patterns of production, labour and consumption."[156] Lifestyle and logistics may unfold in the same physical spaces of the city, but they occupy vastly different worlds—on the one hand a world of instant gratification and on the other, the final stretch of a commodity chain that starts with the logistics revolution and ends in the Uber City.

But unlike the relatively pristine aisles of the supermarket or warehouse, the metropolis is messy. Navigating the city's chaotic rhythms and uneven surfaces adds another layer of work to an already thankless gig. Digital maps and wayfinding instructions streamline the information needed to complete a ride or make a delivery, but they also abstract and flatten the city into undifferentiated, measurable, and manipulable dimensions. As we heard from workers, those abstractions can be entirely deceiving and incoherent; and as workers are increasingly left to their own devices, they must develop their own sense of the "worthwhile" that recognizes cities as neither flat nor abstract. It is up to workers to sort out the awkward integration of what geographer Martin Danyluk calls logistics' "spatial fungibility" and "geographic specificity."[157] Any residual costs of this stubborn commensuration are left to workers to incur; meanwhile, platforms and their investors collect the dividends.

Workers' willingness to align their "calculations" with the platform's—to act "rationally"—will depend on who you ask. Most workers I spoke to came from middle- or upper-middle-class backgrounds. Several were in college or had just graduated. Others held administrative positions at universities or medical centers. Harry, the Uber driver, recently finished med school at an Ivy League institution. Others treated "gigging" as a side hustle to support entrepreneurial efforts such as freelance web design or photography. As platforms ramp up their deceptive and exploitative managerial tactics, these relatively well-off workers are not likely to stick around for the long haul. Workers in my socioeconomic bracket will finish college or med school, land a residency, perhaps get a promotion, but they will not plan a career around gig work. Like all markets, the platform economy shapes its workers into subjects, and this subjectivitization introduces a morality that takes flexibility, independence, and autonomy as its benchmarks.[158] Platforms promised to rewrite the social contract of work but then immediately began chipping away at it. I will be the first to admit that I was happy to hang up my orange Caviar backpack when my stint as a courier ended. And as middle-class workers like myself become disillusioned, platforms are turning to traditionally marginalized groups—racial and ethnic minorities, immigrants, people without college degrees and without another career to fall back on when the on-demand bubble is finished popping.[159] While workers from all social classes may experience work's declining moral economy, it will only be the well-off that have the means to walk away. As Nunberg put it, "The idea of a gig is only alluring if you know you can hit the road when it gets joyless."[160]

If platforms represent the vanguard of economic life in the smart city, then their operations on the ground look nothing like the innovation utopia of the Uber City that Ratti and Claudel foresaw. We need governance models that take worker protections and employer accountability seriously, not as impediments to "innovation." Because if "innovation" is thriving, then what it looks like for the platform workforce is entirely regressive: precarious, fickle, instrumental, extractive. Instead of disaggregating the firm and eliminating the need for internal planning and control, we instead see a complex interplay of coercion and free choice, outsourcing and internalization—a rewiring of work's social circuitry, a science of management through flexibility. All these developments proceed according to terms and conditions set entirely by the firm. And as the platforms' terms of service suggest, all you have to do is accept.

3

Predict

The Logistics of Police Reform

> The police maintain superiority not simply through a monopoly on the use of violence, but by creating a monopoly on the use of logistical media.
>
> —JOSHUA REEVES and JEREMY PACKER, "Police Media: The Governance of Territory, Speed, and Communication"

> Predictive policing is just another form of supply-chain efficiency.
>
> —INGRID BURRINGTON, "What Amazon Taught the Cops"

A Techno-Political Gambit

The day that Ferguson police officer Darren Wilson shot and killed unarmed eighteen-year-old Michael Brown in the middle of Canfield Drive, it was hot out. A call had gone out on the police band to be on the lookout for two young Black men, and Wilson came upon two individuals matching the description. When Wilson ordered them to move to the sidewalk from his SUV, Brown approached the vehicle, reportedly with his hands outstretched. That was when Wilson opened fire.[1]

Such incidents rarely garner much attention. In the news cycle, the shooting was one in a long string of police killings of unarmed Black men, women, and children that would continue after Brown's death—Trayvon Martin, Eric Garner, Eric Harris, Walter Scott, Freddie Gray, Sandra Bland, Laquan McDonald, Tamir Rice. It was the aftermath of Brown's death that exacerbated already boiling tensions and that propelled Ferguson—the tiny suburb of St. Louis in the heavily segregated St. Louis County that Brown called home—onto the national stage as a symbol of police brutality and racial division. Due to a combination of neglect and complicated jurisdictional overlaps, Brown's body lay on the pavement in the broiling August heat, without cover, for hours; when the body was eventually removed, pools of blood stained the baking concrete.[2] The police's handling of the corpse duly outraged the community, and on the evening of Brown's death, August 9, 2014, Ferguson residents gathered at the scene in a show of solidarity, calling for justice and decrying the violence. When officers later destroyed a makeshift memorial of flowers and candles that Brown's mother had helped construct, the vigil took a turn for the worse, with looting and property damage to ensue.[3] The next day, peaceful protests led by local religious figures were surrounded by police in riot gear.[4] Images of these and other dramatic scenes circulated widely, first on social media, then on the national and international news.[5] Weeks of protests followed in Ferguson and in neighboring municipalities, and at each confrontation, police came armored and armed with machine guns, some driving small tanks.

It is difficult to parse Michael Brown's killing from the aggressive response to the protests that followed. Both reflect profound problems that have plagued policing since the establishment of the first law enforcement institutions—the police as a hostile, antagonistic, and intimidating force, sometimes alien to the communities they purport to serve. The demographics of Ferguson and other majority-Black municipalities in St. Louis County, for example, hardly reflect the predominantly white constitution of the police departments that serve them; some officers live in distant rural areas across state lines, maintaining both a geographic and social distance from the districts they patrol.[6] Such discrepancies are not new. As sociologist Alex Vitale argues in his abolitionist critique of policing, modern law enforcement in the United States cannot be divorced from the histories of slavery, colonialism, and worker suppression.[7] The problem is far bigger than the events in Ferguson. Zeke Edwards of the American Civil Liberties Union likewise argues that Ferguson is not the "anomaly" that some pundits made it out to be:

> Ferguson doesn't exist outside the United States; it is the United States. And similarly, our criminal justice system does not exist outside of Ferguson; it's

> an extension of Ferguson.... Our criminal justice system is also an important extension of our legacy, a legacy that goes ... from slavery to forced labor, to Jim Crow to mass incarceration. There's a lot of talk about implicit bias, about unintended effects. But in fact, all of those things, including the criminal justice system, were not unintended and they weren't implicit.[8]

American law enforcement has always been a project of profiling, of race- and class-based exclusions and repressions. The events in Ferguson were merely a clarion reminder of that history or, at the very least, a wake-up call to anyone who thought the past dead.[9]

What has changed, however, are the tools and techniques that police agencies use in their efforts to reform, professionalize, and bolster their legitimacy in the face of discrimination and corruption charges. Law enforcement organizations have long turned to new technologies in an effort to rein in deadly officer discretion and abuse, to make police agencies more accountable to the publics they serve, to rationalize and standardize law enforcement tactics and strategizing—the who, where, and how police officers operate.[10] But technology now enjoys a particular esteem among police intellectuals as a means of effecting meaningful reforms. Indeed, an entire industry of "police tech" now flourishes based on this promise, with offerings ranging from enhanced records management systems and cloud storage to body-worn cameras, predictive algorithms, and crime-fighting dashboards—and with big name investors to boot, including Amazon CEO Jeff Bezos, actor Ashton Kutcher, and former general and CIA director David Petraeus.[11] Police leaders and tech evangelists alike regularly invoke the tech industry's "solutions" as the answer to criminal justice's most stubborn and pernicious injustices—racial bias, overly aggressive tactics, military-style enforcement, military-grade equipment, and so on. And when the cops are bad, business is good. In the week after police officers shot and killed Alton Sterling in Baton Rouge, Louisiana, police tech giant Taser (now Axon) saw its stock jump by 17 percent; its rival, DigitalAlly, enjoyed a nearly 70 percent bump.[12]

In the aftermath of Brown's death, technology certainly seemed to be the answer for the St. Louis County Police Department (SLCPD). The agency was struggling to improve its image. Although Wilson, the cop who shot and killed Brown, was a sworn officer of the Ferguson Police

Department (FPD), the protests that followed were policed by the neighboring SLCPD, which, along with the FPD, was roundly criticized by the U.S. Department of Justice (DOJ) for its response to the unrest.[13] It was SLCPD officers who showed up armored with heavy artillery to accost unarmed protestors, that called in SWAT teams and fired beanbag rounds and tear gas at protestors, striking television crews and Missouri state senator Maria Chappelle-Nadal in the process.[14] The department's belligerent response even sparked condemnations from other police leaders. The commissioner of the St. Louis Police Department (the city, not the county) refused to send additional officers to help contain the protests. All told, the unrest left dozens injured among both protesters and the police. Over 320 people were arrested, including state senator Jamilah Nasheed.[15]

It was only shortly after all this social and political upheaval (not to mention an eventual DOJ investigation into the Ferguson and St. Louis County police departments) that the SLCPD brought in product specialists and senior analysts from HunchLab for a consultation. HunchLab is a "predictive policing" software developed by Philadelphia-based spatial analytics and software firm Azavea. To some, the phrase "predictive policing" may evoke scenes from the 2002 Steven Spielberg film *Minority Report*. Based on a Philip K. Dick short story of the same name, the film depicts a future in which a trio of mutated humans with near-impeccable premonitions report crimes to police before they are ever committed. Outside of sci-fi, of course, crime prediction does not rely on psychics, and predictive policing cannot actually predict the future. But that has not stopped criminal justice institutions from purchasing expensive tools from tech vendors. Today, predictive policing designates a wide range of applications. Where some systems specialize in offender-based modeling (for instance, Chicago's Strategic Subjects List or the risk assessment algorithms used by courts to predict recidivism risk among defendants),[16] other products focus on data fusion across criminal justice agencies (as in the software developed by secretive CIA-backed Silicon Valley firm Palantir for U.S. Immigration and Customs Enforcement to track and deport undocumented immigrants).[17]

HunchLab, by contrast, is a place-based prediction tool, designed specifically for municipal policing. Its software uses advanced machine learning algorithms, trained on years' worth of historical crime data, geographic information, and other data sets, to predict the location and timing of future crimes and then make patrol resource-allocation recommendations accordingly.[18] Even within the niche of geospatial predictive policing applications, HunchLab faces competition. Some place-based prediction models are developed, piloted, and sold by academic criminologists (as in the Risk Terrain Modeling program run out of Rutgers University), while others are produced by major tech companies (Hitachi's Predictive Crime Analytics or Motorola's CommandCentral Predictive, for example). Others still, such as PredPol, are backed by venture-capital funding.[19]

All this competition meant that the team at Azavea had to distinguish HunchLab from its rivals. HunchLab promised greater sophistication, access, and transparency—all qualities that would resonate with departments like the SLCPD, whose reputation and operational autonomy were on the line. Indeed, the HunchLab team has worked in some capacity with the New York, Chicago, and Philadelphia police departments (the first-, second-, and sixth-largest city police agencies in the United States, respectively), each pursuing predictive policing in the wake of corruption allegations or scandal. And with videos of sworn SLCPD officers cursing and aiming automatic weapons at unarmed protestors now circulating on social media and in the news, St. Louis County likewise needed to shore up its image.[20] HunchLab would be the centerpiece of the department's reputation-building campaign, the cornerstone of its public demonstration of commitment to professionalization after months of bad headlines.

In the previous chapters, I explored the economics motivating smart city intervention: in chapter 1, we saw how public–private partnerships to fund urban infrastructure prioritize the logistics of ad circulation over information access and privacy; in chapter 2, we examined how platforms leverage managerial control over the flexible workforce to transform the city into a space of logistical value extraction and seizure. Here I take a different tack. While much could certainly be said about the booming business of police tech and the ways that financial interests distort public safety and technology policy, the questions I address here have less to do with profit outright than with technology's merits and shortcomings as a solution to policing's ongoing crisis of legitimacy. What happens when the tools and techniques of the smart city are held up as answers to the criminal justice system's violence? How did optimal resource allocations and productivity hacks become a commonsense solution to police bias and discrimination?

The idea that predictive analytics might advance fairness, accountability, and efficiency finds support among both police leaders and civilian advocates—computer scientists, technology vendors, policymakers. Early proponents argued that algorithms would be "force multipliers" for cash-strapped departments—a way to "do more with less."[21] The first mumblings of predictive policing came at the height of the Great Recession, with already fiscally strapped agencies facing the prospect of further cutbacks. In 2009, Charlie Beck, then Chief of Detectives and later Chief of Police for the Los Angeles Police Department, coauthored an article in *Police Chief* magazine challenging law enforcement leaders to learn from retail giants like Amazon and Walmart—companies that were using algorithms to "anticipate or predict future demand." Police departments, Beck argued, needed to do the same: "New tools designed to increase the effective use of police resources could make every agency more efficient, regardless of the availability of resources. . . . As the law enforcement community increasingly is asked to do more with less, predictive policing represents an opportunity to prevent crime and respond more effectively, while optimizing increasingly scarce or limited resources, including personnel."[22]

Beyond these efficiency boosts, advocates also made an ethical case for predictive policing. More than merely optimizing resource or personnel allocations, new technologies had the potential to "make the legal system fairer." For example, in a blog post for *Scientific American*—"How to Fight Bias with Predictive Policing"—data scientist Eric Siegel describes systems like HunchLab as "an unprecedented opportunity for racial justice" and an "ideal platform on which new practices for racial equity may be systematically and widely deployed."[23] A *New York Times* opinion piece reached similar conclusions. And while the authors of the *Times* piece—academic computer scientists and computational social scientists—acknowledged the need for checks and balances to avoid disparate outcomes, they ultimately conclude that "well-designed algorithms can counter the biases and inconsistencies of unaided human judgments and help ensure equitable outcomes for all."[24] These laudable goals are not likely to be lost on police department executives, especially at a time when only 16 percent of the public view police officers as protectors of the peace, and when a full two-thirds of officers believe that police killings of Black Americans are "isolated incidents," not indicative of a broader pattern.[25]

Other researchers reach starkly different conclusions. Against claims of fairness and accountability, a growing number of empirical investigations have found policing algorithms to be profoundly unfair and unaccountable—reifications of societal biases, discriminatory by design or otherwise.[26] Julia Angwin and her colleagues at *ProPublica,* for instance, showed that COMPAS—an algorithm used by courts across the United States to make bail determinations—was more likely to incorrectly flag African American defendants as "high risk" than white defendants, and more likely to incorrectly flag white defendants as "low risk." More broadly, the idea that automated decision-making systems inherit the biases of their makers, or even society at large, now appears regularly in mainstream news.[27] In the context of police tech, the critiques have generally been leveled at the data used to train predictive algorithms. In 2016, the ACLU issued a statement of concern, cosigned by multiple civil liberties organizations, suggesting that "the data driving predictive enforcement activities—such as the location and timing of previously reported crimes, or patterns of community- and officer-initiated 911 calls—is profoundly limited and biased."[28] Algorithms trained on data that are limited, incomplete, inaccurate, and skewed due to decades of discriminatory police practices will only reinforce the disparate treatment that marginalized communities already expect from police agencies. At the same time, the algorithms neglect, or at best overlook, what sociologists and criminologists have long known to be the most significant causes of crime: structural racism, systemic disinvestment, and poverty.[29] The fear is that such systems create vicious feedback loops that validate cycles of over-policing—a process that Rashida Richardson, Jason Schultz, and Kate Crawford of the AI Now Institute argue, starts and ends with "dirty data," the digital traces of police abuse and discrimination.[30]

Predictive policing thus sits at the center of what can seem like an intractable debate. How can the algorithm be both a *source of* and *solution to* police abuse of power? How does a patrol technology like HunchLab become a corrective to the legitimacy crises from which departments like St. Louis County hoped to recover?

These questions get at the ethical limits of predictive policing as an actually existing smart city intervention.[31] While the police tech industry has its own momentum, we also see crime prediction featured prominently in mainstream smart city campaigns. Tech giants routinely include crime forecasting and analytics in urban dashboards and smart city software. IBM's Smarter Cities products, for instance, include crime prediction tools ("technology solutions for crime prevention") alongside healthcare and education solutions.[32] On the one hand, this suggests a certain flattening, a sense that policing can be "optimized" in the same way that transportation or sanitation might be—or, as Beck suggested, in the same way that Amazon or Walmart streamline their supply chains. With a more rational and efficient use of police resources, the logic goes, police would have the latitude to return to fundamentals—maintaining order and deterring crime. On the other hand, with agencies like the SLCPD turning to predictive policing in the midst of crisis, it is clear that policing's peculiarity as an institution requires a nuance that has all too easily been overlooked by predictive policing's champions. The police do not simply distribute public safety benefits more or less efficiently. They operate at the interface between a city's populace and the state's monopoly on the legitimate use of violence; and to members of marginalized communities who have suffered at the brute end of that equation, policing itself becomes a harm. Predictive policing operates within this ambiguity, and it does so, like programmatic advertising and the platform economy, based on questions of optimal distributions and allocations, of the spaces and times of policing as an infrastructure—but without ever resolving the tensions.

This chapter investigates the techno-political gambit of predictive policing as a tool for police reform, a gambit whose assumptions I argue are deeply flawed. That machine learning and automated decision-making tools could be framed as the solution to police discrimination fundamentally misunderstands what the technology is capable of and what the problems are that plague police departments. Like earlier efforts to reform law enforcement with new technology, predictive policing operates on the *logistics* of police patrols. It does not dismantle the patrol; it reformats the patrol along the axes of time and space to suit managerial imperatives. Squad cars, two-way radios, mobile data terminals, and, more recently, body-worn cameras—each of these tools has been imagined to work similarly, extending law enforcement's surveillant gaze into the polis while expanding managerial control over officers in the field.[33] A critical understanding of predictive policing means grappling with this dual gaze. And it must do so while appreciating the ways that the algorithm *integrates* with existing patrol technologies. As critical technology scholar Lucas Introna argues, algorithms "must be understood in situated practices—as part of the heterogeneous sociomaterial assemblages within which they are embedded."[34] Predictive policing embeds the algorithm within the police patrol—a contradictory enactment of the state's power to distribute both public safety benefits and harm across uneven geographies of authority and legitimacy, risk and danger, that the police themselves help construct and maintain.[35]

The chapter builds on ethnographic research conducted with the product team at HunchLab. Rather than studying with any of the dozens of police departments currently using predictive policing, I instead focus on how Azavea's product managers and programmers talked about the technology, how they pitched the product as an ethical alternative in the crowded marketplace for police tech, and how they framed and designed its algorithmics to suit the mandates of police reform. This focus was in part influenced by the company culture at Azavea—a firm that proudly celebrates its certification as a "B Corp," or "social benefit" company, meaning that it refuses to accept venture financing and explicitly prioritizes ideals of transparency and openness. Researching with the HunchLab team was therefore fortuitous, not only because Azavea's transparency commitments got me in the door as an ethnographer in an industry known for its furtiveness, but because HunchLab offers a glimpse onto the ethical frontiers of technology design in a context where the stakes of fairness, accountability, and transparency are so high. How did the HunchLab team imagine its algorithmic interventions making policing fairer, more accountable, and more transparent? What are the limits to those efforts? With what frictions will they be met?

Two Views of Patrol Technology

Throughout the history of modern policing, new technologies have appeared as correlates, if not catalysts, to organizational change in law enforcement. From the police callbox, two-way radio, and squad car in the first half of the twentieth century, to the mobile data terminals, records management systems, and computer-aided dispatch of the latter half, the historical trajectory of modern urban policing is punctuated by its evolving material, technical, and infrastructural supports.[36] Most of these developments are now so deeply woven into the fabric of policing that they appear as natural extensions of the patrol officer.[37]

Predictive policing and other recent developments—body-worn cameras, gunshot detection systems, automated license plate readers, facial recognition—extend this lineage into a new era of data-driven "intelligence." Most of these products integrate with well-established surveillance and patrol practices—the beat, the uniform, the prominently placed marked squad car. Crime prediction algorithms, for instance, change little about the everyday mechanics of policing other than where officers are sent on their patrol shifts. By design, the algorithm is not visible. In some jurisdictions, communities would have no idea that local police are using crime prediction tools were it not for the efforts of investigative journalists, activists, and researchers bringing that information to light. The Brennan Center for Justice at New York University, for example, had to sue the City of New York to get information about the NYPD's secretive predictive policing trials.[38] In other cases, even politicians and government officials have been left in the dark about police contracts with tech firms. In San Francisco, the department went behind the backs of the civilian oversight body to discuss a potential contract with geospatial predictive policing firm PredPol.[39] In New Orleans, Silicon Valley data analytics giant Palantir worked with the police for years without the City Council's knowledge.

All of which is to say, in their guise as tools of reform, most new police technologies fly under the radar. They reconfigure and augment existing operations, but without upending or dismantling them. As critical media and race scholar Lyndsey Beutin argues, the tools of reform "incrementally improve, but ultimately maintain, the existing system."[40] And it is precisely because of this incrementalism that police leaders so often invoke new technology in the wake of controversy or scandal—to save face without giving up authority. This has been true historically, as we will see below, but it is especially true for the current crop of data-driven decision-making systems. As critical legal scholar Andrew Guthrie Ferguson writes, the adoption of big data policing strategies, including predictive policing, "grew out of crisis and a need to turn the page on scandals that revealed systemic problems with policing tactics."[41] Consider the $6 million predictive policing experiment underway in Chicago. As part of broader reform efforts, the Chicago Police Department (CPD) has been trialing a host of technical fixes, including an algorithm used since 2012 to predict perpetrators and victims of gun violence.[42] But these efforts expanded dramatically after former mayor Rahm Emanuel colluded with the CPD to block the release of dashcam video showing officer Jason Van Dyke shooting and killing unarmed Black teenager Laquan McDonald. Massive protests and a DOJ investigation followed when this information came to light, and although Van Dyke was later found guilty on charges of second-degree murder, Emanuel's approval rating dropped to the low twenties.[43] "In response to demonstrated human bias," Ferguson writes, "it is not surprising that the lure of objective-seeming, data-driven policing might be tempting."[44] The cycle of police violence, protests and unrest, and DOJ investigation now appears to conclude with city police agencies adopting big data solutions as a fix to institutional discrimination, abuse, and injustice.[45]

Institutional Narratives of Innovation and Reform

The association of new technologies with police reform is consistent with institutional narratives of police culture—stories that law enforcement institutions tell about themselves. Predictive policing is one of those stories. To advocates, predictive policing is the "next phase" in a technology-driven march toward better policing, the culmination of a series of technological and organizational innovations meant to professionalize police agencies and improve their relationship to the communities they "serve."[46] To understand why predictive policing finds such strong support among its proponents, we need to dive into these institutional narratives and consider where crime prediction fits into the larger story of the police's "evolving" relationship to technology.

A good starting point is "The Evolving Strategy of Policing," an influential text authored by two leading police intellectuals, George Kelling of the conservative Manhattan Institute and Mark H. Moore of Harvard's Kennedy School of Government. The article, published in *Perspectives on*

Policing—a prestigious joint initiative of the DOJ, the National Institute of Justice, the Police Foundation, and the Kennedy School (i.e., the most influential institutions in U.S. policing)—traces the history of law enforcement's institutional culture and organizational mandate.[47] Although technology is not the obvious protagonist of the story, it is clear throughout the text that when police agencies decide to adopt new technologies, they do so as part of a broader shift in the priorities and logics of policing. New investigatory techniques, information infrastructures, transportation tools, and communication networks all present as rational, if imperfect, responses to fluctuating obligations in the police's function, political authorization, tactics, and methods.

In this sense, Kelling and Moore's text offers nothing less than an institutionally vetted story of technology's role in institutional change—the police's internal mythology. And according to this mythology, police history maps onto three periods—the political, reform, and community problem-solving eras. The political era lasted from about the 1840s to the 1930s and coincided with the commissioning of most municipal police forces in the United States and Europe. During this period, urban policing was a nascent and largely corrupt institution, commissioned by local political bosses, corporations, or large property owners. In the northern half of the United States, early police departments were an outgrowth of private or informal security structures (such as the Pinkertons or town watches) and were generally concerned with bullying votes for political parties; subduing ethnic, religious, or racial strife during periods of mass immigration; or quelling worker dissent.[48] In southern states, police forces formed from slave patrols. In port cities like Charleston and New Orleans, police subjected people of color (enslaved and free Blacks alike) to constant monitoring, inspection, and everyday violence—conditions that evolved after the U.S. Civil War into racist legal codes forcing African Americans into political and economic subservience.[49]

By the 1920s and '30s, these institutions had become a stain on the legitimacy of the political bodies they represented. Police corruption was so rampant that pressure to reform came from multiple directions—business leaders, journalists, religious leaders, and, eventually, politicians. In 1929, President Herbert Hoover appointed the Wickersham Commission to investigate law enforcement's ineffectual response to organised crime under Prohibition.[50] The investigation bolstered efforts among leading police intellectuals, such as Berkeley, California, Chief of Police August Vollmer and New York City Police Commissioner Theodore Roosevelt, who had launched comprehensive professionalization campaigns urging departments across the country to rationalize their procedures through new investigatory standards and patrol technologies.[51] Throughout the "reform era," agencies responded to these calls with bureaucratic and technocratic interventions—civil service exams, crime labs, fingerprinting, new management techniques, mapping, new communications and transportation infrastructures, and so on—all meant to professionalize police forces and enshrine scientific ideals of objectivity in their procedures.

Kelling and Moore generally attribute these developments to Vollmer, whose influential work on police administration was particularly appealing to politicians and police executives eager to clean up their unpopular forces. But Vollmer and his disciples were also driven by what they perceived to be the country's growing moral deficits. As Kelling and Moore write, "Police in the post-flapper generation were to remind American citizens and institutions of the moral vision that had made America great and of their responsibilities to maintain that vision."[52] Of course, the morality of any call to "make America great again" is limited by race, class, and gender inequities, and the police reformers' "morality" was equally confined to white, heteronormative, middle- and upper-class values. As a consequence—and despite its objective veneer—policing in the "reform" era turned out to be no less caustic for poor and marginalized communities than during the political era. To racial and sexual minorities and political dissidents (such as gay or lesbian communities or Communist or Socialist party members),[53] the façade of rigor and objectivity cultivated by Vollmer, Roosevelt, and their followers was merely the professional face of violent, occupation-like policies and procedures. In 1950s Los Angeles, for example, the "reformed" police department's mandate seemed to be merely enforcing racial segregation. This earned the agency a reputation for its harsh discrimination of Black and Latinx residents in the postwar years. LAPD Chief of Police William Parker, who served from 1950 until his death in 1966, was largely credited with "professionalizing" the agency through rigorous, military-like discipline. But as civil rights lawyer Connie Rice argues in the powerful documentary *Burn*

Motherfucker, Burn! Parker's reforms "didn't get rid of the racism, and so the racism took on an occupying-force quality."[54]

The mid-twentieth-century reforms may therefore have "rationalized" and systematized police procedures, but they continued to be applied within oppressive, racist regimes.[55] As Beutin contends, "If the system of policing is inherently racialized and racializing, strengthening those institutions, even in the name of reforming them, will not address the underlying structural issues."[56]

In part, a recognition of racial politics drove the transition to the next phase in policing's history, the "community problem-solving era." But racial politics alone did not usher in the transition. According to Kelling and Moore's history, the reform era would last into the 1970s, after a convergence of forces culminated in the political turmoil of the 1960s—civil rights activism, antiwar protests, televised images of violent police tactics, rising crime rates, and a general disenchantment among both the public and police officers with the myths of "objective" and "impartial" law enforcement. As Kelling and Moore put it, the reform strategies of the mid-twentieth century were simply "unable to adjust to the changing social circumstances of the 1960s and 1970s."[57] With civil and racial unrest, school integration, white flight, and the early seeds of deindustrialization and job shortages all disturbing the postwar idyll, the police had been tasked with cleaning up the mess by using the only tools they knew—rational, systematic, but no less violent tactics.

Urban racial and economic landscapes were undergoing massive change during this period, and community activists, criminologists, and police leaders alike began questioning key tenets of the reform era's prioritization of routinization and rationalization. With capital fleeing the city and crime spiking, the rapid response protocols and displays of police authority that had been enshrined as reform-era ideals betrayed a growing distance between the police and the public. Officers were not serving communities—much less collaborating with community members—so much as shoring up and reinforcing their own authority. Calls for change even came from line officers themselves, who had grown increasingly disillusioned with the professionalization efforts. Police management may have been spouting rhetoric of one's duty "to serve and protect," but the brass continued to treat street cops as low-status and their work as routinized and standardized—for instance, by enforcing petty internal rules governing officers' appearance and off-duty activity.[58]

As these concerns mounted, criminologists began urging police administrators to jettison the reform era's top-down crime-fighting approaches. They recommended tactics that could be tailored to local crime "problems" identified by police in collaboration with community leaders and criminologists. Indeed, the rise of community problem-solving policing tracked with a wholesale revolution in the field of criminology—what critical sociologist David Garland calls a "transformation of criminological thought."[59] Starting in the 1970s, criminology exploded with field studies, experiments, and trials, all intent on empirically vetting the police's "tried-and-true" tactics. This new criminology would abandon the field's traditional focus on the criminal psyche and instead investigate the social and physical contexts in which crimes are committed—crime "problems," or what Garland calls "criminogenic situations."[60] What street lighting or neighborhood layouts correlate with problem areas or "crime hot spots," for instance? As Garland writes:

> The assumption is that criminal actions will routinely occur if controls are absent and attractive targets are available, whether or not the individuals have a "criminal disposition" (which, where it does exist, is in any case difficult to change).... The new policy advice is to concentrate on substituting prevention for cure, reducing the supply of opportunities, increasing situational and social controls, and modifying everyday routines. The welfare of deprived social groups, or the needs of maladjusted individuals, are less central to this way of thinking.[61]

The new criminology thus "normalized" crime, in the sense that crime was no longer seen as the product of the criminal subject's abnormal constitution, but an everyday social reality that could be mitigated more or less successfully.[62] If police reformers imagined that criminal deviance could be wholly eliminated through rationalization and standardization, community problem-solving would concede crime's stubborn persistence and experiment with tailored approaches to local crime problems. According to Kelling and Moore, this meant giving officers more discretion, since "discretion is the *sine qua non* of problem-solving policing." Officers needed the autonomy to identify and triage problems, to determine which issues demanded the most attention and, consequently, the discretion to act without prohibitive procedures: "Problem solving is hardly the routinized and standardized patrol modality that reformers thought was necessary to maintain control of police and limit their discretion."[63] If the reformers' "model of police as impartial, professional law enforcers was attractive because it minimized the discretionary excesses which developed during the political era,"[64] then they also allowed these directives to get in the way of the police's most fundamental mandate: upholding the "social order" at all costs.[65] The problem-solving ethos clawed back a heroic authority for the street cop walking her beat, "emphasiz[ing] a wider definition of the police function and the desired effects of police work."[66] and enshrining order-maintenance not only as the outcome toward which police–community partnerships should strive, but as a normative metric of the police's success.

The transitions between each of the three eras in Kelling and Moore's institutional mythology correspond with the implicit promise and then perceived failure of technological overture. The reform era responded to corruption in the political era by establishing new professional standards, ideals of scientific objectivity, and information and communication infrastructures—the national crime reporting database was started in this period, for example, and nearly every department adopted standardized dispatching systems to coordinate increasingly auto-mobilized patrol fleets. The community problem-solving era, in turn, was a rejoinder to the miscarriage of these developments. Where the reform era emphasized routine, standardization, and professionalized procedure to rein in officer discretion, community problem-solving policing demanded officer flexibility, autonomy, and context-specific controls. Its answer was a new "intellectual technology"[67]—a new theory and analytics of crime that would be oriented to the spatial and temporal determinants of criminal activity as well as the effectiveness of police responses to that activity. At each turn, technology is imagined to adjust the police's course to realign with the public interest.

Kelling and Moore penned their account in 1988. Subsequent analyses extend the institutional narrative into a fourth phase in the twenty-first century—into what criminologist Dennis Rosenbaum calls the era of "information technology" and what researchers at the RAND Corporation describe as "intelligence-led policing."[68] Regardless of the designation, "better and faster technologies" are now widely seen to have ushered in a new epoch in policing's institutional evolution.[69]

Compstat, the managerial overhaul of the NYPD pioneered by Deputy Commissioner Jack Maple at the behest of Commissioner William Bratton in the 1990s, was a watershed for this information-intensive era.[70] A portmanteau of "computerized" or "comparative statistics" (apparently none of Compstat's historians can agree on this point), Compstat introduced data analytics as the crux of organizational reform. Crime maps and statistics would now be the centerpiece of special weekly or biweekly meetings, where department executives grilled local precinct commanders over their jurisdiction's crime numbers. If crime was on the rise in a particular district, executives raised the issue at Compstat meetings, probed for explanations, and held the district's commanders responsible to higher arrest rates or crime reductions—often using crude language or, famously, by throwing chairs to communicate the gravity of the situation.[71]

Under Compstat, crime data came to be seen as an infallible metric of trickle-down, crime-fighting accountability. With Bratton and Maple evangelizing their managerial system across the United States and around the world, policing became an information business.[72] Crime rates started falling in New York and other U.S. cities through the 1990s and into the 2000s, and many credited Compstat with the decline. Others, however, charged that Compstat created perverse incentives to "juke the stats," to falsify data and manipulate numbers—what criminologists John Eterno and Eli Silverman called the "crime numbers game."[73]

Regardless of its efficacy or fallibility, Compstat was undeniably associated with a wide-ranging transformation in policing commonly known as "order-maintenance" or "broken windows" policing—a model based on a concept that Kelling himself helped make famous.[74] Broken windows was a contradictory model of policing.[75] On the one hand, following from Compstat's obsession with crime numbers, broken windows policing introduced search and arrest quotas, including the now-infamous "stop and frisk" policy (which Michael Bloomberg's administration expanded). These quotas extended department executives' managerial authority down to line officers' activity in the field, limiting their discretion to decide which low-level offenses deserved policing and which could be dispensed with a stern warning. On the other hand, and in line with Kelling and Moore's argument for greater officer discretion, broken windows–style policing gave line officers broad latitude to determine what constituted a threat to the "social order" and "quality of life"—and therefore who constituted a criminal.[76] And with the passage of the Clinton administration's 1994 crime bill, the crackdown on the streets was matched in the courts by mandatory minimum sentencing. While police were enjoying the autonomy to criminalize low-level crimes like turnstile-jumping and public intoxication, judges were left with far less discretion to adjudicate punishment for victimless crimes.[77]

The consequences of these developments have been severe: the number of people incarcerated in U.S. state and federal prisons nearly doubled between 1992 and 2017.[78] But to leading police intellectuals like Bratton, the problem with Compstat was not that it produced perverse incentives through its quotas or that it helped feed the rapidly growing carceral state. The problem was that Compstat remained largely unrefined. Its retrospective analytics were based on rudimentary "hot spots" of crime. And it offered little nuance in dealing with crime problems other than brute interventions and intensive patrols. A more refined tool was needed.

If Compstat ushered in law enforcement's "information technology" or "intelligence-led" era, *predictive policing* would be its technological corrective, the "next phase" in the evolving strategy of policing.[79] And indeed, Bratton is generally credited with coining the phrase "predictive policing" during his tenure as LAPD chief, as well as with playing an outsized role in its development.[80] Where Compstat was a "police performance measurement system,"[81] predictive policing would be about proactive deterrence; where Compstat came up short operationally, predictive policing would leverage even more powerful statistical techniques and increased computational speed—"better and faster technologies"—to bypass opportunities for human error or misconduct.

Early conceptions of predictive policing were simple, if vague, and perhaps overly ambitious. In 2009, when Charlie Beck (then working under Bratton at the LAPD) called on police leaders to adapt techniques from e-commerce, his articulation of predictive policing's "strategic foundation" was relatively straightforward—"a smaller, more agile force can effectively counter larger numbers by leveraging intelligence."[82] But what this meant in practical terms was fuzzy. During a pair of symposiums organized by the National Institute of Justice in 2009 and 2010, attendees broke into workgroups to discuss what, exactly, "predictive policing" could mean operationally. Responses varied. Some suggested that predictions might be used for riot or crowd control; others recommended tools for sex-offender risk assessments, police-personnel management, and prisoner-radicalization prevention. Ultimately, the symposium's participants were able to give a unified definition that was broad enough to encompass nearly any application of data analytics to policing—"any policing strategy or tactic that develops and uses information and advanced analysis to inform forward-thinking crime prevention."[83]

With such a broad mandate, predictive policing's possibilities seemed limitless. But as private contractors, criminologists, and crime analysts translated those embryonic ideas into concrete software applications and platforms, a more precise picture of predictive policing began to emerge: advanced algorithms, trained on historical crime data, to predict the locations or perpetrators of future criminal activity, and thereby prevent the crime from taking place. Where community problem-solving demanded greater discretion for officers to identify crime problems, predictive policing would automate that discretion, vetting decisions on the street to software-sorted assessments of crime's spatial and temporal determinants.[84]

The Contradictions of "Policing by Consent"

In the institutional narrative, predictive policing represents a refinement on refinements—the apex of technical and organizational innovation in law enforcement, a return to policing's fundamentals of crime prevention and deterrence. These ambitions are on full display, for example, in a televised commercial for IBM's Smarter Cities Crime Prediction and Prevention tools. The ad opens at night, with a police officer checking his wristwatch and getting into his patrol car. Off somewhere else, a shady-looking character also checks the time and drives off. The two men zoom across the city, at one point passing each other unknowingly. Meanwhile, a convenience store clerk is bagging the day's cash for deposit, an easy mark for a stickup. The officer narrates as a clock ticks away—*I used to think my job was all about arrests, chasing bad guys. Now I see my work differently.* He checks the squad car's mobile data terminal as he drives—*We analyze crime data, spot patterns, and figure out where to send patrols.* The shady-looking character arrives at the convenience store, putting on gloves as he approaches the door—*It's helped some U.S. cities cut serious crime by up to 30 percent*—but the cop is already there—*by stopping it before it happens.* Defeated, the would-be criminal turns away.[85]

This magic-like preemptive power is precisely how Charlie Beck foresaw police analytics working in the midst of recession, and it is how Bratton continues to talk about predictive policing—a more efficient allocation of scarce resources, a more "precise" form of policing.[86] In this regard, the ad perfectly captures the hopes for the technology articulated in institutional narratives: crime predictions so accurate that the cop gets to the scene before a crime is ever committed. But as we will see, this scenario is as fictional as *Minority Report.* Any claim to crime predictions' accuracy is impossible to substantiate. Furthermore, the scenario depicted in the IBM ad boosts expectations for the technology, creating a false sense of surety and hype around predictive policing that, the HunchLab team felt, must be constantly tampered.

On a more fundamental level, however, the institutional narrative is inadequate not only because it exaggerates the precision or accuracy of crime predictions, but because it overlooks the contradictory mandates of modern policing. The officer in the commercial may have arrived to the scene of the crime in time to preempt the would-be offender, but police officers in the real world routinely flaunt their authority in violent displays: in 2015, North Charleston, South Carolina, officer Michael Slager shot unarmed Walter Scott in the back during a pursuit that followed a routine traffic stop for a non-functioning brake light.[87] We need another framing, another view of patrol technologies, one that refuses to take crime prediction for granted as the culmination of a progressive march toward superior policing through "better and faster" technology, that recognizes policing's ambiguous role in liberal governance as both "respectful protection" and "intrusive penetration."[88]

Crime *prevention* (rather than detection) has long been a pipe dream of modern police institutions. In the IBM commercial, predictive policing figures crime deterrence as an artifact of the technology, the predictive algorithm. This apparent novelty obscures the deep historical basis for prevention and deterrence in the earliest formulations of—and justifications for—the "science of police."[89] In the eighteenth and nineteenth centuries, foundational police intellectuals such as Patrick Colquhoun or Edwin Chadwick argued explicitly that police must act as a "preventative force whose detective work is secondary."[90] When the British Parliament passed the London Metropolitan Police Act in 1829, prevention was enshrined in the agency's founding charter. The first of Sir Robert Peel's famous "Nine Principles of Policing," issued at Scotland Yard's commission, states that the "basic mission for which the police exist is to *prevent* crime and disorder." It is no coincidence that William Bratton refers nostalgically to predictive policing as a technological revival of Peel's principles.[91]

But if crime prediction algorithms return policing to these mandates, then we must also take seriously the contradictions baked into the model of "liberal-consent policing" that Peel's principles embodied.[92] According to social theorists Markus Dubber and Mariana Valverde, early police bodies served a "hinge" function for the burgeoning liberal state. In the abstract, the concept of police linked otherwise incommensurable and temporally disjointed logics of juridical power—the backward-looking criminal law, which punishes crimes ex post facto, and the future-oriented logics of prevention, anticipation, and preemption that intervene before any crime has been committed. "Prevention and punishment," Dubber and Valverde write, "are very different logics of governance; 'police' is the middle term that links them."[93] This linkage becomes particularly dangerous for marginalized populations when the state's abstract power to police is translated into actually existing police forces. Prevention endorses a preemptive authority to intervene—to police—before crimes are ever committed, even if that authority repudiates the police's democratic commission and mandate to "serve and protect" unobtrusively.[94] Despite the ambitious liberalism espoused in Scotland Yard's charter, Peel's laissez-faire principles quickly gave way to a form of localized sovereignty in the figure of the "bobby," the constable on patrol.[95] The discretion to determine who needs protecting and who needs policing—those who constitute the polity and those who represent a threat to its order—has been the officer's prerogative since the establishment of the first modern police agencies.

When viewed from this wide angle, predictive policing does not signal a technologically enabled realignment of policing's mandate with shifting public expectations, nor is it the latest phase in "intelligence-led policing." Instead it represents a reassertion of the police's contradictory functions, a fine-tuning of the tensions that have always defined modern police—as *a distribution of both public safety benefits and state-sanctioned violence.* In practice, "maintaining the peace" has always meant targeting repressive interdictions on those at the peripheries of society while excluding poor and negatively racialized communities from police protections. In the United States and other colonial contexts, for example, the spectacle of terror against enslaved bodies served a deterrent function by "making an example" of even minor displays of disobedience through violence.[96] We see the legacy effects of this "deterrence" in broken windows and order-maintenance-style policing, where police crackdowns and carceral punishment are used to deter victimless, quality-of-life transgressions.[97]

Such historical parallels are invisible to the institutional narrative because it fundamentally misrecognizes the relationship between technology and the most elemental of policing practices—the beat, the patrol, the neighborhood watch. In the institutional narrative, technology is something that *happens to* the patrol; the patrol, meanwhile, remains outside the technological domain. Reform-era standardization rationalized the patrol's procedures; Compstat's crime data increased patrol accountability through quotas. But this says nothing about the patrol itself as a *mediating technology,* one that activates in the roving distribution and inscription of police officers in and through the urban landscape.

To claim that police patrols are a mediating technology requires a radical theory of mediation. Following anthropologist William Mazzarella and media theorists Sarah Kember and Joanna Zylinska, media and mediation are not ancillary social functions—they are constitutive of social and political reality.[98] Social life is never premediated or immediate. "Mediation does not serve as a translational or transparent layer or intermediary between independently existing entities," Kember and Zylinska argue. It *produces* entities, and gives the appearance of independent existence; it brings entities into being as social forces and sociomaterial configurations.[99] "Mediation," Mazzarella likewise suggests, "is a name that we might give to the processes by which a given social dispensation"—the city, the nation, politics, religion, economy, crime, safety, risk, and so on—"produces and reproduces itself in and through a particular set of media."[100] Mediation makes society imaginable to its members through distancing, optics, and objectification, and it is in this expansive, radical sense of mediation that I mean the patrol mediates: it produces "crime" and "safety" through its mediation of the urban landscape.

The patrol enjoys a special status as the elemental mediation of the modern state's civil power—the material and technical framework through which the abstract power to police touches down on the ground through actually existing police forces. Whether it is the constable walking his beat, the officer driving her route, or the trooper monitoring the highways, the patrol is a "reflexive and reifying" technology, "inseparable from the movement of social life and yet removed from it . . . at once obvious and strange, indispensable and uncanny, intimate and distant."[101] If police are the hinge between prevention and punishment, then the patrol binds the power–knowledge of statistics and cartography to the sovereign authority to intervene—to stop, search, and seize; to make move.[102] As historian Willem De Lint writes, the patrol operationalizes law enforcement "as a question of the allocation of men to territorialized (or spatialized) jurisdiction."[103] Its spatial and temporal techniques mobilize officers as an infrastructure of preventive enforcement, a circulation of the law's gaze, with resources allocated "based upon the mobility demands of the fleeing criminal and the communication imperatives of a responsive/preventive police apparatus."[104] The patrol enacts a spatial economy to support its distribution of public safety benefits and violence across geographies of risk and hazard that it constructs through its mobilizations.[105] According to philosopher Jacques Rancière, the power *to police* is the power to partition the social world, and this is always an ambivalent act—a separation and exclusion that also enables participation and inclusion.[106] We might therefore imagine the *patrol* as the ritualized enactment of the urban partition, a mediation of the city's uneven geography of risk and danger, legitimacy and authority, that "produces a fiction of its premediated existence."[107] As certain populations, behaviors, or features are sanctioned, others are marked as suspicious, deserving of inspection, intervention, or violence.[108] And through the ritualized operation of the patrol, these mediations appear as natural contours of the city's social topology.

Political theorist Thomas Nail offers a detailed accounting of the patrol's mediating functions.[109] First, the patrol is *preventive* and *circulatory*—it deters crimes before they take can place by "oscillating its presence to and fro."[110] Patrols do not police crime per se. They police the criminogenic and policeable landscape as a surface of movement. They not only stop and inspect (and search and seize), they also *make move*—they conduct and "keep moving," smoothing circulation while providing "dromological support" for their own efficient movement. Second, the patrol activates what Nail calls the *kinoptic* function—a watchful circulation and spectacle of surveillance, a constable that *sees* and *is seen* in the same instance, making her presence known as she monitors the scene. Ideally, the patrol's circulations are crafted "as a moving image of perfection and order"—or what Chadwick described as an "ambulating lighthouse."[111] The image of the lighthouse is especially evocative of the patrol's kinoptics: it illuminates and makes visible while at the same time making its own presence known. And the entire apparatus is enacted according to rational schema—the *kinographic* function, the patrol's inscription of movement and geography, its mapping of urban movement. Through kinographic inscription, the patrol rationalizes the most efficient routes and routines for its own double optic, mapping criminal potentialities and optimizing its preventive circulations—all of which requires a sprawling cartographic and documentary assemblage to synchronize the patrol as a mesh of ubiquitous presence in time and space.

With these mediating functions, new patrol technologies no longer appear as correctives to institutional crises of legitimacy, as Kelling and Moore's "evolution" suggests. They are tools for expanding the patrol's authority over an urban landscape that the police itself constructs "as a technological and infrastructural problem [of] how best to organize and regulate flows of people, commodities, and risks."[112] As critical communications scholars Joshua Reeves and Jeremy Packer argue, with new patrol technologies police expand their authority "not simply through a monopoly on the use of violence, but by creating a monopoly on the use of *logistical media*."[113]

Such logistical monopolies are ambivalent, however. In addition to expanding police presence across time and space, new technologies invite opportunities for what De Lint calls "supervisory co-presence."[114] The same tools and techniques that amplify logistical control also connect officers in the field to their superiors, subjecting them to managerial exposure and scrutiny. "With each new technology," De Lint writes, "a fuller and more penetrating gaze has been envisioned, *both of the police into the polity and of police supervision on police officer mobilization.*" Patrol technologies "structure the decision making of individual officers on patrol to organizationally vetted formats."[115] On the one hand, such increased supervision may be framed as a response to public pressure to "professionalize" the police force. Better supervision checks officers' discretion and promises to rein in misconduct and abuse. In this sense, we find some overlap with the institutional narrative, with new technologies presented as accountability measures. On the other hand, tightened supervision must also be understood as a solution to officer autonomy as a *managerial* problem. According to criminologist Lawrence Sherman, police discretion is a perennial concern in police management: "No matter what targets are selected for police resources, no matter how well the police methods are tested, the central management question will always be, 'what are police doing to accomplish our objectives, when, where, and with what apparent result?'"[116] Supervisory co-presence in this sense is part of the kinographic assemblage. It ensures that the patrol's allocations to territorialized jurisdiction are *optimal* distributions, with officers positioned most efficiently to respond to situations demanding their attention. And while new technologies' gaze may trouble line officers as infringements on privacy or autonomy, to management they are entirely necessary to assure the patrol's functioning as a collective organ—as a fleet, as an infrastructure, as an organization and regulation of flows.

When we view the patrol itself as a sociomaterial mediation, it becomes difficult to imagine predictive policing as a refinement on earlier generations of technologically enabled patrol. Like other patrol technologies, it represents a double-edged expansion of surveillance and managerial control, both into the polity and onto the police. IBM's analytics may promise to anticipate crime more efficiently, but they also introduce new forms of logistical governance that graft onto the patrol assemblage (the squad car, the two-way radio, the mobile data terminal, and so on). This duality helps explain why predictive policing is so controversial yet so readily adopted; how proponents can argue that the algorithm mitigates police harms while critics insist that it exacerbates biased and discriminatory policing. Predictive policing rationalizes the surveillant spectacle of the patrol—the police as "ambulating lighthouse"—while fostering an algorithmic supervisory co-presence, tightening managerial directives for where, when, and how patrol officers circulate. The attraction is that predictive policing might finally resolve the fundamental incommensurability at the heart of urban policing—the impossible balance of "respectful protection" and "intrusive penetration."

Predictive Policing for Reform

Predictive policing therefore appeals to police departments hoping to extend accountability, not to district commanders (as was Compstat's prerogative) but to line officers in the field. As artist and researcher Ingrid Burrington argues, while predictive policing may have little effect on what police officers actually *do* when out on patrol (other than where they go), "it does have the potential to increase the power that police management has over cops on the street."[117] This is predictive policing's appeal. It comports with current trends in applied criminology toward "evidence-based policing," where managerial tools like GPS and body-worn cameras track "what police were or were not doing in relation to the dynamic patterns of crime and public safety problems,"[118] while refining the crude crime analytics driving targeted patrols. It extends the institutional gaze onto both "the cops" and "the robbers" through supervisory co-presence and kinographically rationalized preventive circulations.

If most critiques of predictive policing center on the police's expanded gaze, they leave unresolved the question of how expanded managerial control will be operationalized.[119] Will predictive policing reprise the reform era's proceduralism of standards, routines, and professionalization? Will it allow police greater discretion to solve "crime problems"? Will it be used in tandem with order-maintenance, broken windows–style tactics? As Burrington writes, the DOJ investigation into the Ferguson and St. Louis County police departments showed that "the tendency toward micromanagement too often leads to more petty arrests in pursuit of revenue and quotas."[120] Whether predictive policing remedies this tendency or creates new perverse incentives is difficult to answer, as most police agencies remain tightlipped about whether and how they actually use the technology. We can, however, gain some insight by looking to the algorithm's design, to the ways that the technology's producers imagine crime prediction to reconfigure and reformat the patrol. How do algorithm designers envision end users—the patrol officers—engaging with predictive "intelligence"? How do they imagine the mutual imbrication of "intuition" and "data," of human hunches and machinic allocations—and how does this entanglement translate to a less-harmful model of policing?

HunchLab

The findings are based on ethnographic research with the HunchLab product team at Azavea, a company that specializes in web-based geographic data applications—"advanced geospatial technology and research for civic and social impact."[121] From October 2015 to May 2016, I visited the company's offices in Philadelphia's trendy warehouse district between shifts as a platform courier. I participated in business meetings, sat in on planning sessions, gave feedback on webinars, met with potential clients, traveled on site visits, and attended all-staff events, both with the HunchLab team and Azavea's full staff. Although the company has over fifty employees, the HunchLab team itself was small. Jeremy Heffner, HunchLab's product manager and senior data scientist, led a group that at any given point consisted of only two to three full-time product specialists. These were recent college grads placed at Azavea through a fellowship with Venture for America, a program that matches entrepreneurial hopefuls with startups. And while these product specialists managed sales and marketing, they did not do the coding or software-writing for HunchLab. These tasks were completed by one of Azavea's multiple programmer teams, which work on several projects and products concurrently. In addition to participant observation with HunchLab team members and extensive interviews with Heffner, I interviewed programmers, Azavea founder and CEO Robert Cheetham, and several criminologists and police officers that HunchLab hired intermittently as consultants.

Within the field of predictive policing, HunchLab enjoyed a unique status coming out of Azavea. Most competitors are brought to market by large corporations (IBM, Microsoft, Motorola, Hitachi, LexisNexis, and so on), often as a component of holistic smart city dashboards. Other products are sold by smaller, venture-capital-backed companies, such as PredPol, or by firms started up with seed funding from the CIA, such as Palantir.[122] Azavea, by contrast, is not beholden to shareholders, investors, or covert government funding schemes. Instead it adheres to a strict set of criteria for corporate social responsibility, environmental sustainability, and transparency—all of which have earned the firm a certification as a "social benefit" company, or B Corp. Cheetham, for example, encourages employees to devote a certain portion of work time to pro bono, "socially impactful" projects; each year, Azavea sponsors a Summer of Maps fellowship that pairs undergraduate data science students with mentors to complete projects for nonprofit organizations.

For the HunchLab team, Azavea's social-good commitments translated into outreach. As HunchLab's senior manager, Heffner regularly participated on panels alongside prominent critics of surveillance and police technology, for instance, at the Brennan Center for Justice at New York University. And while these commitments distinguished HunchLab within the field of police tech, within Azavea, the controversial nature of predictive policing made the product an awkward fit with the company's vibe. In-house discussions and debates over the ethics of predictive policing occurred regularly, and on more than one occasion I observed the product team turn down expansion opportunities or partnerships due to ethical concerns.[123] And although HunchLab's mission states explicitly that it is designed to reduce the harm associated with over-policing, at least one programmer chose to opt out of any HunchLab-related work—a request that the company honored.

The first HunchLab prototype was built with grant money awarded by the U.S. government through the NSF.[124] Cheetham had previously worked as a civilian crime analyst with the Philadelphia Police Department (PPD), where he developed a technique to measure crime spikes. After founding Azavea, Cheetham was approached by his former supervisor who suggested that they apply together for the NSF grant.[125] Cheetham agreed and brought on a team of Temple University criminologists to help evaluate the proposal, hiring Heffner to manage the project.[126] Despite Cheetham's experience with the PPD, it was Heffner, a mathematician and statistician, who introduced techniques from artificial intelligence and machine learning to the new HunchLab prototype. With these advanced methods, the team could avoid having to commit to any particular crime-forecasting approach. Given fast enough computing and processing power, multiple prediction methods could be incorporated into a single, theory-agnostic meta-model, capable of "representing concepts from several different approaches."[127] These included: (1) the crime-spike detection system that Cheetham developed for the PPD; (2) a variation of "near repeat" analysis, a forecasting technique pioneered by UK criminologists to predict the spatial and temporal distribution of crimes, and particularly home burglaries;[128] and (3) risk terrain modeling, which uses the proximity of crime events to urban features (bars, churches, transportation hubs, and so on) to generate geospatial risk profiles.[129] HunchLab's model incorporates each of these methods, operationalizing criminological concepts as modular parameters and variables.

When I conducted the research, HunchLab was sold as a subscription service. Pricing was determined by the size of a client jurisdiction's population, but it generally started at about $50,000 for the first year and $35,000 for subsequent years.[130] Clients gained access to several features, but the algorithm driving its crime predictions was called Predictive Missions. Like other crime prediction tools, Predictive Missions is trained on a client's historical crime data—when possible, up to five years' worth. But unlike other systems, HunchLab combines crime data with as much non-crime-related information as possible—census data; weather patterns; the locations of bars, restaurants, and bus stops; moon cycles; school schedules; holidays; concerts and events calendars—all, Heffner and Cheetham argued, with the intention of mitigating the bias baked into crime data that are managed by the police themselves. Once aggregated, the information is mapped onto a grid of five-hundred-square-foot cells laid over a map of the client's jurisdiction. A series of thousands of decision trees recursively partitions the data based on crime outcomes in each grid cell. If a crime occurred in a cell, the regressions determine the variables that influenced the occasion of that crime and to what extent. This establishes how closely each variable correlates with each type of crime and then weights the variables accordingly.

The result is a hyper-localized and hyper-sensitive forecasting algorithm that tailors crime predictions to each client's jurisdiction by crime type (see Figure 14). Because the algorithm adjusts weights according to

Figure 14. The HunchLab user interface shows a map of the client's jurisdiction, with grid cells shaded by color to indicate risk for different crime types. The map here shows a cross section of North Philadelphia, made up of predominantly African American neighborhoods, some of which have crime rates higher than the city's average. Image courtesy of ShotSpotter.

signals in the local crime data, models for the same crime will differ across jurisdictions; and because the data are updated daily or even between shifts, the weighting for each crime type may change to match shifting patterns in criminal behavior. For example, *location* might be the most predictive factor for *theft from automobiles* in Detroit, but in Philadelphia it could be *time of day*; both models automatically adjust if the crime patterns changed. When the modeling was evaluated against ground truth data (that is, where crimes actually took place), the results indicated high levels of accuracy—sometimes as high as 92 to 97 percent, depending on the crime type.[131]

Indeterminacies

The HunchLab team advertised these accuracy rates to potential clients. This was the case on a clear and cold December morning in St. Louis

County, as I huddled with Heffner outside one of the SLCPD's nondescript operations centers waiting to meet his contact in the department. Heffner told me that he had run accuracy tests on the St. Louis County crime data the night before. The results were better than usual, and Heffner was excited. Within the hour, he would be presenting HunchLab to the SLCPD brass at a Compstat meeting and would pull up those accuracy tests on his slide deck to demonstrate the algorithm's precision.

Although St. Louis County had not even started using HunchLab, that did not prohibit Heffner from running the accuracy tests. In fact, the HunchLab product specialists regularly created mock-up models for cities or jurisdictions that were not clients but which publicly released their crime data as part of open governance and civic tech programs.[132] Accuracy tests simply required withholding recent crime events from the training data and then comparing those to the predictions. An accuracy rate of 92 percent, which was the case for St. Louis County, meant that for every hundred crime predictions, only eight were false positives or otherwise incorrect; competitors, by contrast, report accuracy rates at around 20 percent, at the highest; more often, they report in the single digits.[133] Ninety-two percent was worth bringing up.

Even with these incredibly high accuracy rates, however, the HunchLab team acknowledged a fundamental limitation to their forecasts: *predictive accuracy can only be measured prior to implementation.* Once a department starts putting algorithmic predictions into practice and officers begin patrolling predicted crime locations, the ground truth data ceases to be a controlled sample. Officers' visible presence in a predicted location (their kinoptics) changes what takes place there, and thus the conditions being modeled; at the same time, the officer's presence produces new (kinographic) data—if she makes an arrest or responds to a call for service in a predicted crime area, that activity shows up in the data and is fed into the models for the next day, even the next shift. These two effects pull in opposite directions and therefore confound accuracy measurements. They invalidate comparison between the predictions and actual crime observations. Any evaluation of predictive accuracy after implementation thus becomes meaningless. Like the Heisenberg uncertainty principle or post-humanist theories of performativity, HunchLab imagined the double optics of the patrol's kinoptic and kinographic functions—the observable and inscribable position of officers in the field—as introducing statistical indeterminacies at the same time that they produce new data.[134]

Heffner conceptualized these effects as a paradox animated by competing probabilities: *detection*, the increased likelihood that an officer observes a crime taking place at the predicted location; and *deterrence*, the increased likelihood that the officer's visible presence prevents crime from taking place there.[135] As Heffner explained:

> If we give [the police] HunchLab mission areas across the entire city, and they start putting officers in those mission areas throughout the city, at that point we can no longer accurately measure our ability to select the right locations. Right? Because if they're going to those locations, they're going to have some effect in terms of detecting new events—because they're there—and deterring events—because they're there. And those things work in opposite directions. So, it's very hard to have a clean measurement of accuracy once you start using the output.[136]

The effect is an epistemological impossibility—an indeterminacy that is characteristic of algorithmic prediction in general but largely absent from debates around big data policing.[137] As sociologist Adrian Mackenzie argues, prediction introduces two distinct sets of difficulties, one concerning predictive *performance* (how accurate the predictions are), the other concerning *performativity* (the effect that acting on predictions has on the conditions being modeled).[138] With the detection–deterrence paradox, predictions' performative effects confound performance measures. The failure to appreciate this indeterminacy stems from a widespread and inaccurate theorization of predictive policing as the work of an invisible and disembodied statistical apparatus, detached from the sociomaterial assemblage of the police patrol.[139] But once this dynamic is recognized, the task becomes to capture, measure, and analyze the predictions' effects on the world *within the modeling*—as Mackenzie puts it, by "fold[ing] the performativity of models back into the modelling process."[140]

In predictive policing, such "folding" can work if the desired outcomes are measurable behaviors or actions. If clients want predictions to produce higher arrest rates, this can be modeled because arrests are observable in the data. But if the desired outcome is *prevention*—as in Sir Robert Peel's final principle that the ultimate "test of police efficiency" must be "the absence of crime and disorder, not the visible evidence of police action in dealing with it"[141]—then system designers are faced with another epistemological impossibility: an event deterred is by definition unobservable, and thus immeasurable (as in the truism that you cannot prove a negative). Of course, prevention rates may be inferred by comparison between a treatment group and a control group.[142] But even this will always be an imperfect estimation, as no two jurisdictions, beats, or patrol shifts are ever identical. Further, maintaining a control group necessarily means only partially implementing predictions, a prospect with which most clients will not be happy given that they are paying handsomely for a subscription.

Ultimately, the detection–deterrence paradox introduces an indeterminacy to crime prediction that cannot be avoided once put into practice, and the HunchLab team appears to be alone among police technology vendors in recognizing this difficulty. But as we will see, the team also hoped to use that same indeterminacy as an opportunity for intervention. Indeterminacy opens a space for predicted outcomes to be thwarted and deterred, to rewrite the geography of public safety benefits more equitably, to reduce the harms of over-policing—even if we can no longer measure the predictions' accuracy. When that which is statistically unobservable (crime prevention) is the most desirable and least harmful outcome, indeterminacy must be an acceptable condition.

Prescriptive (rather than predictive) analytics is the name that the HunchLab team gave to their attempt to *work with* statistical indeterminacy. In a webinar titled "Beyond the Box: Towards Prescriptive Analysis in Policing," Heffner and product specialist Chip Koziara explained the rationale. In business analytics, prescription occupies a more complex register than prediction.[143] Rather than merely predicting an outcome based on past events, prescription seeks to account, first, for the effects that predictive information has on the decision-making itself—that is, what Mackenzie calls performativity—and second, the effects that those decisions have on future outcomes—how to exploit that performativity to achieve desirable results by identifying and optimizing trade-offs.[144] "Beyond just making a prediction," Heffner explained, it is about "actually giving a prescription or some sort of guidance about decisions that would be most effective to make in a particular situation to effect the result that you want to have occur."[145]

We see these logics applied in HarmStat, a program that Heffner was developing with Temple University criminologist Jerry Ratcliffe during my time at Azavea.[146] HarmStat starts from the progressive notion that police presence in low-income and minority neighborhoods is less likely to be perceived as a public safety benefit than a source of harm in its own right—and that the harm of over-policing can be quantified and evaluated against the harms of crime victimization. In essence it asks, which is worse—over-policing or under-policing?

At root, HarmStat operationalizes that question as a cost–benefit analysis: it estimates the "costs" of police harms and compares these to the costs

Crime Models

Label	Severity Weight	Patrol Efficacy	Patrol Weight	Relative Weight
Theft from Vehicle	2,100	75%	1,575.0	1.0
Residential Burglary	13,100	24%	3,144.0	2.0
Homicide	8,649,200	1%	86,492.0	54.9
Motor Vehicle Theft	9,100	75%	6,825.0	4.3
Robberies	67,300	10%	6,730.0	4.3
Aggravated Assault	87,200	10%	8,720.0	5.5

+New Crime Model

Figure 15. This screenshot from the HunchLab interface illustrates the quantification involved in "harm-focused" policing: "severity weight" is derived from FBI calculations of "cost per crime," "patrol efficacy" gives an estimate of the crime's preventability, and "patrol weight" is the product of the two; clients can adjust "relative weight" to indicate the importance of preventing the most harmful crime relative to the least harmful. Here, *homicide* is configured as 54.9 times more important to prevent than *theft from vehicle.* Image courtesy of ShotSpotter.

of crime harms (based on the FBI's Uniform Crime Report, which attributes a dollar amount to crimes) to make a determination for police intervention. The trouble is, HarmStat uses an estimate called "predictive efficacy" that imports and then buries indeterminacy deep within the analysis. Predictive efficacy indexes a crime's predictability in relation to both its preventability and its potential impact on the community—in other words, how likely are we to successfully predict the crime, how likely are we to successfully prevent it from taking place, and how important is this prevention to the community? For example, by nature, violent crimes like assault and homicide are less predictable than property crimes like burglary or larceny. But communities are also more likely to place a high premium on preventing violent crimes because they cause more harm.[147] Conversely, while efforts to thwart the more easily predictable crimes may have greater success rates, these attempts might not be as important to community members and may therefore be read as overpolicing. HarmStat (and HunchLab's modeling generally) is designed such that these preferences are configurable within the system as a series of weighting mechanisms (see Figure 15). Heffner explains:

> The missions that we generate include the different types of crime that the police department wants to address, as well as the importance of preventing those crimes and how effective[ly] they believe each crime to be prevented via patrol. So, for example, something like a homicide would be much more important to prevent than a robbery, but a homicide is very unlikely to be preventable. And so, you can adjust the configuration within the system to reflect this.
>
> Now, let's say that even given those configuration options, you still get a homicide location [appearing] on the map. . . . So, how does that inform what you'd do? Your initial reaction might be "Well, we really can't prevent homicides," and that's true. But you're representing that [unpreventability] within the configuration of the system, right? So, maybe you say that you only have a 1 percent likelihood of preventing a homicide but yet a homicide box still showed up. And it's showing up here in relation to other choices. The system could choose another location [with] a different type of crime. . . . But yet it's choosing *this* location. And so that can inform a policy decision. . . . In such unusual situations, where you have a

> low-volume crime-type that's not preventable but still shows up as a focus, [it] might actually warrant [sending] resources when you typically wouldn't think to.[148]

Because the system is configured to reflect each crime's predictability, preventability, and severity, when the homicide box "still shows up," it means that the algorithm has flagged something of significance—that despite homicide's *un*predictability, the risk is great enough to warrant a response.

The idea makes sense intuitively, and it reflects HunchLab's goals of optimizing patrol resources to prevent harmful outcomes.[149] But the logic also glosses over the indeterminacy baked into the harm modeling. First, if Predictive Missions is already being used to allocate resources, then it is impossible for HunchLab or their clients to evaluate predictive accuracy, meaning that the "predictability" metrics will be based on only immeasurable assumptions of precision. Second, and perhaps more troublingly, predictive efficacy relies on an estimation of deterrence (or "preventability") that is ultimately unobservable. Aside from the entirely unlikely scenario that police observe a crime deterred (as in the IBM commercial), it is impossible to know whether sending an officer to a particular intersection prevented a murder from taking place there. And if we consider the faith that police are likely to place in the predictions, these epistemological loopholes can become harmful.[150]

We can imagine how this might play out. Homicide predictions are liable to put the police on high alert, prompting district commanders to amplify the police's presence in the predicted crime location. Perhaps the cops come in plain clothes, perhaps they bring in the tactical division. Regardless, what officers *do* with that "intelligence" at the predicted location remains entirely at their discretion. The police could resort to the kind of quality-of-life policing that targets low-level offenses with harsh enforcement; they could pursue "suspicious-looking" individuals in the area; even their mere presence or contact with community members in the neighborhood could increase anxiety symptoms.[151] All of these possibilities may or may not result in a homicide prevented—we can never say for sure.

I observed this dynamic in action the day after Heffner presented HunchLab to the SLCPD brass when I went on a ride along with an officer using HunchLab for the first time. The algorithm directed where the officer should patrol but did nothing to prevent him from drawing on previous experience, training, and intuition once we arrived in a grid cell. At one point, he spotted a rental car on the perimeter of an area predicted to be at high risk for larceny. The rental-car driver had done nothing wrong, but the officer explained to me that rental cars can be a sign of suspicious activity. When the car turned at an intersection without signaling, the officer used the offense as a pretense for a traffic stop; this then turned into a vehicle search when the officer smelled marijuana in the car. Although none of this had anything to do with a larceny, our presence may or may not have deterred the occasion.

HarmStat's "predictive efficacy" metrics reflect an analytic gamble: that the harms of over-policing can be outweighed by the public safety benefits of a crime deterred. But whether that deterrence can be credited to the prediction is unknowable, and this may allow policing to proceed as usual, unreformed. With statistical indeterminacy embedded at multiple points in the decision-making structure, the basic configuration of "predictive efficacy" rests on sociotechnical faith and police intuition.

Trade-Offs

Predictive Missions is at the root of each HunchLab function—the algorithm constantly adjusting crime models and generating risk scores for each jurisdiction's thousands of grid cells. But it is the Allocation Engine algorithm that analyzes these thousands of cells to select locations for patrol. The Allocation Engine is essentially a sorting function, a process that all predictive policing systems use to parse information as "intelligence." But unlike other systems, the Allocation Engine's defining feature is the selective and strategic insertion of *randomization* into the sorting process. Rather than directing patrol to grid cells with the highest risk—which is how HunchLab was originally designed and how its competitors operate—the algorithm instead sends officers to the second, third, fourth, fifth (and so on) riskiest areas based on a probabilistic but randomized selection process.[152]

The HunchLab team invoked randomization's benefits often and strategically, both in their pitch to potential clients (for example, in meetings with crime analysts or police commanders) and when facing criticism from civil liberties activists or researchers. But they also acknowledged that randomization requires a sacrifice. Randomizing the sorting and selection process means that even if the predictions *are* highly accurate (which cannot be measured), the cells selected for patrol may not be. The HunchLab team viewed this compromise as productive—if you cannot measure accuracy, why not sacrifice it for other benefits? "If we're just trying to maximize our predictive accuracy," Heffner clarified, "then, absolutely, selecting that [highest risk] cell every time would be what we'd do. But that's not the case here. You're going to act on [these predictions] and so you're going to start to skew things, displace crime, and so forth."[153] Randomization makes sense, in other words, because predictive policing is not only about predicting crime; it is about managing the logistics of police control.[154]

When the HunchLab team addressed potential clients, they described several benefits associated with randomization. First, they argued that randomization would make patrols less predictable to potential offenders and thus alleviate the well-known criminological concern over crime *displacement.* This occurs when targeted or "hot spot" patrols do not actually prevent or deter crime but simply push criminal activity into a new area.[155] And because many observers view predictive policing as a high-tech upgrade to hot spot policing, its application has renewed anxiety over displacement effects.[156] HunchLab responded to these anxieties through webinars and self-published reports, explaining that the Allocation Engine's randomized cell selection would prevent high-crime areas from being "oversaturated" by patrol, making officer circulations and routes less predictable to potential offenders and thus less likely to push crime "around the corner."[157]

Second, the team invoked randomization as a means of combating officer boredom—a way to keep officers on their toes. Although officer boredom may seem like a trivial issue, criminologists have found that boredom can lead to adverse outcomes. As Scott Phillips writes in his ethnography of police work, "When a person has nothing to do, he or she will find something to do, and that something may not be positive behavior."[158] The issue came up frequently among police officers. "In Seattle," one consulting police officer explained, "they had to shelve [a predictive policing system] within a handful of months because the officers had no idea what to do, and they were saying, 'I'm bored.'"[159] A survey of the Burbank Police Department likewise found that algorithmic patrols produced a malaise among line officers, leaving them disillusioned and ultimately disinterested in the system.[160] The risk is that predictive policing appears, as criminologist Dave Allen has put it, as merely a flashy new face on the same old "augmentation" tools that feel like "management by remote control."[161] And if this is what predictive policing feels like, then officers are not likely to take the crime predictions seriously. Algorithms are useless if the cops do not bother going to predicted crime areas, which is precisely what sociologist Sarah Brayne observed in her study of predictive policing in the LAPD.[162]

What HunchLab needed, then, was "buy-in." Buy-in is always important, criminologist Jerry Ratcliffe explained to me, drawing on his expertise as both academic and former police officer. But buy-in is especially important in large and "more loosely coupled" departments like Philadelphia's, where the command staff has little control over officers' day-to-day activities.[163] "You have to get buy-in because [police are] less responsive to direct orders. There just isn't the capacity or the will there to have really close supervision. But that means you have to get people's buy-in, because otherwise they'll just do their own thing."[164] So while HunchLab had a reputation for wonky demonstrations and sophisticated tech (which made the platform popular among crime analysts and department executives holding advanced degrees), its sophistication was unlikely to appeal to line officers, especially if the software was perceived as an exercise in de-skilling. Officers needed to figure out for themselves that HunchLab was "credible," as one officer put it, by helping them make an arrest—or what he called a "touchdown": "If your software could say, 'Okay, we've seen this type of crime before and this is the tactic that worked,' do you know what type of credibility you would get in any department you launched this in?"[165] HunchLab product specialist Adele Zhang concurred, suggesting that the best situation is when word spreads "organically" among the rank and file: "If an officer has some sort of good experience, feels that they did something, and it's because of HunchLab, they're going to be like, 'Oh, yeah, I did this—and it worked.' And it only needs to be anecdotes. It doesn't need to be real from an overall perspective, it just needs to be an anecdote."[166]

Of course, word-of-mouth approval would be great, but it is hard to guarantee. Heffner, for his part, saw randomization as at least a partial answer to the problem. Like the detection–deterrence paradox, he mapped officer buy-in as a trade-off between competing objectives—sending officers to patrol crimes in areas already known to be at high-risk (such as *theft from vehicles* in a parking lot, or *larceny* at a mall) and sending them to patrol areas that were "less obvious"—demonstrating that the algorithm both "knows what it's doing" and is "actually doing something."[167] He explained this to the SLCPD command staff at the Compstat meeting:

> So, your officers might say, "Well, those areas are all areas that we know about." Okay, so, let's say that we didn't give them any areas that they believed to be true, they then wouldn't believe the system, right? And so, conversely, if we only identified areas that they believed to be the highest risk areas, then we wouldn't be adding value. So, what we try to do is a mixture of the two.[168]

Lee Hunt, crime analyst for HunchLab client Greensboro Police Department in North Carolina, found that Heffner's strategy worked. If software vendors wanted the rank and file to get onboard, then they needed to show cops "what they don't know," to make officers think "that the grid cells are somehow special," to shake things up and get officers "off those main thoroughfares and [doing] things slightly differently" by giving them "a different perspective for the same area they're normally patrolling."[169] To Hunt, the Allocation Engine's strategic dosage of randomization created that productive mixture.

By contrast, the civil rights advocates and researchers that the HunchLab team encountered during their outreach efforts were not likely to place much stock in officer buy-in. When Heffner presented HunchLab to these audiences, he framed randomization as an opportunity to intervene into the routinized discrimination in patrol allocations. For example, at a symposium on "Policing and Accountability in the Digital Age" hosted by the Brennan Center, Heffner was joined on a panel by Aliya Rahman, a community organizer and fellow with the progressive New America Foundation's Open Technology Institute and a critic of police technology. During her presentation, Rahman issued a challenge to the audience:

> Is there anybody here who believes that communities marked as more high-risk by [crime prediction] algorithms—is there anyone who believes that black people and people of color are, by nature, riskier, more dangerous people? [Silence from the audience.] Cool, so if an algorithm is saying that this is where [police] should go, then it's broken, correct? Or it is 100 percent performing the wrong predictive task. It is not actually telling us how to be more safe; it's only telling us *how close today can be to yesterday,* which is the mathematical definition of maintaining the status quo.[170]

When Heffner responded, he explained that Azavea was concerned about the same issues, which is why they built randomization into the Allocation Engine:

> If you're a police department and you're not using an algorithm, you're probably using a hot spot map . . . [which] has probably not changed very much for a long time. So, what we do in HunchLab is, we sometimes don't send [patrols] to the highest-risk places, because then we can see what happens when we don't send them there and we send them to a lower risk place. . . . It's a bit of a randomization based upon the analysis to help us gain more insight into what it would look like when you *don't* saturate an area with police. Because maybe we don't have that in the training data and we need to gain that knowledge.[171]

According to Heffner, randomization breaks the cycle; it shuffles the deck and disrupts the status quo. If poor and racial-minority communities are overrepresented in the data due to over-policing, then it is also the case that wealthier and whiter areas are systematically *under*represented. This is what Heffner means by needing to "gain that knowledge."

Heffner's logic here combines two well-known trade-offs in computer science—between fairness and accuracy, and between exploitation and exploration. The fairness–accuracy trade-off describes outcomes when algorithms operate on data about people. Algorithms designed to be "fairness-aware" try to ensure that outcomes do not disproportionately affect members of a protected class (by class, race, ethnicity, sexuality, gender, religion, and so on), but in doing so are forced to make compromises in accuracy—in the same way that randomization sacrifices the predictions' accuracy relative to ground truth data.[172] In the exploitation–exploration trade-off, compromises are made between obtaining new information (exploration) and using the information you already have to improve performance (exploitation).[173] The HunchLab team imagined randomization to *increase fairness through exploration,* in the sense that randomizing patrols means sending officers to areas systematically underrepresented in the data due to policing's entrenched biases (wealthier and whiter areas), while at the same time alleviating the harms of over-policing in communities that are overrepresented (poor and racial-minority areas).[174] If Rahman is right that the algorithm maintains the status quo, then to the HunchLab team randomization was an opportunity to intervene, to clean up the dirty data, to disrupt the "ratchet effects" and "runaway feedback loops" that entrench systemic patterns of bias and discrimination.[175]

While this logic makes sense on the page, it overlooks the extent to which the fairness intervention may conflict with randomization's appeal to police departments. If randomized allocations send officers to the tonier and less-victimized quarters of the city, as Heffner suggests, does this increase the chance that the cops—the system's users—will get bored? If working in wealthier areas means line officers are less likely to get "a touchdown" by making an arrest, will they be less likely to "buy into" the system and grow suspicious of the recommendations? If cops need evidence that the algorithm "knows what it's doing" by allocating resources to areas *already perceived as high-risk,* how will this square with concerns that the algorithm targets communities of color? Any meaningful dent in the dirty data's representational disparities would require a radical reallocation of police resources, likely in ways that undermine the system's appeal to clients—officers—and thus their buy-in. At the end of the day, HunchLab is a product on the market, and in a trade-off of trade-offs, clients' preferences are most likely to prevail.

More fundamentally, at the root of the exchange between Heffner and Rahman are incommensurable notions of "bias." On the one hand, there's the *statistical* bias of systematic over- or underrepresentation of a population in the data; on the other, the *social* bias of racialized populations' historical association with crime and criminality.[176] Randomization may help solve for statistical bias by producing a more "optimal" distribution of police's public safety benefits, but it can do nothing to tamper the social bias—what Khalil Gibran Muhammad calls the "condemnation of blackness"[177]—that sanctions and targets racial violence through the policing of communities of color.

Unfalsifiables

Along with the Predictive Missions crime predictions and the Allocation Engine's cell selection, HunchLab clients also gain access to a software feature called Advisor. Operationalizing the "prescriptive policing" model discussed above, Advisor embeds predictions within an analytic apparatus designed to gather evidence in support of tactical decisions. If the system predicts a *theft from vehicle,* for example, what is the best response to that prediction—what should cops actually *do* in the grid cells? This is the question that experimental criminologists have been asking for decades, and on a superficial reading, Advisor appears to merely automate the randomized control trial—the method of choice in place-based criminology since the 1990s.[178] On a closer read, however, Advisor, like randomization, requires a trade-off—not between accuracy and fairness in this case, but between scientific rigor and "evidence."

Advisor has three initiative types: Field Test, Experiment, and Adaptive Tactics. Field Test evaluates the effectiveness of a given tactic or set of tactics in reducing the incidence of a specific crime. For example, following a wave of *home burglaries,* a department might use Field Test to test the impact of tactics such as *high-visibility patrol* and *canvassing homes and businesses* on the rate of future home-burglary events. To this end, Azavea preloads Field Test with a roster of suggested approaches, such as "writ[ing] reports while parked in patrol cars at high-risk locations."[179] But these fields are also customizable: district commanders or crime analysts can update the entries to reflect whichever tactics they want to test. Once delineated, Field Test compares the incidence of the crime to "what likely would have happened had you not been doing the field test," again based on the Predictive Missions output and again baking indeterminacy into the scaffolding of more advanced analytics.[180]

The second initiative type, Experiment, is similar, but expands the experimental frame to the entire jurisdiction by randomly assigning entire beats, districts, or precincts as control or treatment groups, essentially replicating the methodology of the randomized control trial. But, as Heffner argued, Experiment had certain advantages over academic trials. Because the system is designed for internal testing and not, say, publication in a peer-reviewed journal, clients are less likely to be concerned with achieving proper statistical thresholds for significance. The program is meant for running quick-and-dirty experiments, to see what tactics work best—to gather some evidence before making a decision. With plug-and-chug design and a user-friendly interface, trials can be implemented rapidly, lowering the barriers for evidence-based policing.

Unlike Field Test and Experiment, Adaptive Tactics, the third initiative type, does not confine experimentation to a fixed timeframe. Instead it uses ongoing data collection to examine the effects of a whole panel of tactics on crime outcomes. To start, clients create a list of tactics tailored to each crime type. Predictions of *theft from vehicle,* for instance, might call for a different set of responses than *home burglaries.* Then, each time the Predictive Missions algorithm generates risk scores, Adaptive Tactics makes a corresponding tactical recommendation and the system records the tactic's execution in relation to the risk profile for the grid cell. This begins as a random assignment, but as the system accumulates more data, the recommendations are made with greater inferential confidence. Any correlation between a tactic and deterrence (the crime's absence) serves as an algorithmic "reward" in the machine learning process.

The pragmatic experimentalism across all three of Advisor's initiatives echoes recent calls from criminologists and law enforcement leaders for police departments to become more "iterative," experimental, evidence-based. Between 2014 and 2016, the National Institute of Justice ran a Randomized Control Trial Challenge, offering grants of up to $100,000 for departments to conduct empirical studies of patrol strategies.[181] Jim Bueermann, president of the Police Foundation, predicted that by 2022 every police department would have a resident criminologist running experiments regularly.[182] Criminologist Lawrence Sherman, whose work pioneered the use of randomized control trials in place-based criminological experiments, forecast that by 2025 all major cities' police commanders will be using technologies to conduct routine tests of patrol efficiencies.[183] As a staunch promoter of evidence-driven decision-making, Heffner basically shared these sentiments. But he also noted that criminological studies can be burdensome or overly costly for small departments partnering with academics—at least absent a major grant like that from the National Institute of Justice. "Experimentation and building evidence is [*sic*] an important aspect of policing. . . . [But] is there a way to do this more locally, at home, in a more ongoing fashion?"[184] Heffner reasoned that institutional pressures can be prohibitive for police departments seeking to "iterate" through experimentation. For example, with Compstat's intra-departmental accountability structures still in place, pressures to demonstrate crime reductions "can lead to a sort of risk aversion," making it "less likely to experiment with things. . . . Because if we can just keep things generally as they are, they will likely turn out the same way at the next Compstat meeting—and so that's a safe move."[185] With automation, easy-to-use interface design, and lower thresholds for evidence, Advisor promised to lower those barriers, making it possible for your local police agency to collect evidence and refine their tactical response to crime problems.

And yet, like HarmStat, indeterminate predictions remain the basis for Advisor's tactical evaluations. When Field Test or Adaptive Tactics calculates effects, the system is comparing observed crime outcomes against "what likely would have happened there" had the police not responded to the prediction with the tactic in question. In other words, it is the Predictive Missions outputs—whose accuracy can no longer be verified or evaluated—that provide the "control group." With predictions entangled in more and more of HunchLab's functions, their indeterminacies proliferate. And in a context of experimentation and evidence-gathering, this leads to evidentiary claims that are ultimately nothing more that unfalsifiable inferences—fractions without denominators, control trials with no control group.[186]

The sacrifice of methodological rigor for good-enough estimations is a conceit of big data in general,[187] and in a context as fraught as policing, this should be cause for serious concern. Advisor's experimentalism *could* lead to new, less-discriminatory police tactics; but it could just as easily promote an "optimization" of the practices that led to over-policing, over-criminalization, and over-incarceration of racialized and economically marginal populations in the first place.[188] If predictive policing is a tool of reform, then its success will depend entirely on police agencies' willingness to test less harmful and more equitable responses—canvassing rather than stopping and frisking, flyering rather than calling in the SWAT team. The system determines algorithmic "rewards" for deterrence, but if the only inputs legible to the system are arrests or reports of crime, then Advisor may simply learn how to reproduce the strategies and tactics most in need of change.

As I argued above, police forces have always operated according to an impossible tension between criminalization and prevention, detection and deterrence, punishing and preventing crimes. Calls for reform mount when the former demands overwhelm the latter. We see this continue to play out in St. Louis County. The DOJ investigation into the SLCPD conducted after the Ferguson protests found that the agency had engaged in a disproportionately high number of vehicle stops that led to arrests of racial-minority drivers, and that they did so without taking measures to prohibit racial profiling. As the report states, "While consistent with Missouri data collection law, the traffic stop analysis procedures employed by the SLCPD are inconsistent across the agency and lack the sophistication necessary for appropriate analysis of stop data. This results in a missed opportunity to fully understand if bias-based profiling is occurring."[189] Given the public forum in which the DOJ made these recommendations, the SLCPD was essentially compelled to comply (by public pressure alone, if not legal action). And yet, four years after findings from the DOJ investigation were published, Black drivers continued to be targets of police stops. One study found that in 2018, Black drivers in St. Louis County were still more than twice as likely to be pulled over by police than other racial groups.[190]

Was this post-investigation trend in "driving-while-Black" arrests the SCLPD's response to the DOJ's recommendations? Perhaps a mission to gather more data, more consistently, on the tactic? Or was it simply a failure to comply with reform directives?

It is unlikely that these questions will ever be answered (at least absent another investigation). But with SLCPD now using HunchLab, it is not unreasonable to imagine a system like Advisor being held up as the "solution" to the problem—a way to gather evidence about traffic stops' tactical effectiveness and racial impact. Indeed, even prominent critics of predictive policing have insisted that big data applications could be turned inward, toward officer conduct, to produce what legal scholar Andrew Guthrie Ferguson has called "blue data"—where "the same law enforcement technologies built to track movements, actions, and patterns of criminal activity could also be repurposed to foster data-driven police accountability."[191] After all, collecting traffic stop data more consistently with a tool like Adaptive Tactics would likely satisfy the DOJ's requirements for more and better data for police accountability. And given the extent to which the SLCPD relies on traffic stops as routine police work, it is hard to imagine the agency omitting traffic stops from its software-assisted experimentation.[192]

How, then, would Advisor "optimize" traffic stops? We cannot answer this question without considering the system's sacrifice of rigor for "evidence." Across Advisor's initiatives, tactical "effects" are evaluated against algorithmic crime predictions as evidence of deterrence (whether a predicted crime took place or not), and this opens the door to a particular variety of dirty data—one that is likely to be (statistically) biased toward tactical *effectiveness.* For example, with predictive accuracy no longer measurable because the predictions are already implemented, it is not hard to imagine a scenario in which Predictive Missions *incorrectly* forecasts a crime in close proximity to a traffic stop—like the situation I witnessed on the SLCPD ride along. Because the prediction is incorrect, the crime will never occur. But because we cannot know that it is incorrect, the system would read the outcome as evidence of the traffic stop's "deterrent" effects, and thus issue an algorithmic "reward" that reinforces the tactic's value in the machine learning process. The data cycle would proceed despite the fact that we cannot know whether the crime's absence resulted from the stop or was merely an artifact of inaccurate predictions.

With an experimentalism that sacrifices rigor and falsifiability for quick-and-dirty evidence-gathering, it will only become more difficult to demonstrate disparate impact and, by extension, police profiling, despite the system's apparent satisfaction of DOJ recommendations. Conversely, if we consider the alternative—that the SLCPD's insistence on stopping Black motorists at higher rates than other racial groups was a failure to comply with DOJ recommendations—then it is unlikely to begin with that HunchLab will do anything to "reform" St. Louis County's police. Either way, little changes.

Logistics and the Distribution of Harm and Safety

Predictive policing has always been supported by claims that the algorithm enhances control, certainty, and exactitude. Geospatial predictive policing, in particular, promised to optimize the allocation of patrol resources by leveraging data to anticipate the spatial and temporal determinants of crime. As Charlie Beck put it back in 2009, learning from the predictive logistics of Amazon and Walmart would help cash-strapped departments "prevent, deter, thwart, mitigate, and respond to crime more effectively, ultimately changing public safety outcomes and the associated quality of life for many communities."[193] Far from backing down from these claims, advocates now extend the argument to encompass ethical concerns. When predictions falsely flag Black defendants at higher rates than whites, data scientist Eric Siegel suggests, that information is not incorrect or inaccurate per se; it is simply decontextualized—the algorithm identifies racial disparities in policing, and if we used the analytics to quantify inequality then we could actually do something about it.[194]

The HunchLab case illustrates that Siegel's optimism is no straightforward proposition. When predictive policing's precepts are put into practice, it is not the differences between certainty or uncertainty, precision or imprecision, that determine outcomes or shape decisions; it is indeterminacy that seals predictive policing's fate—the inability to know whether the predictions are inaccurate or not. In theorist Brian Massumi's terms, we might say that predictive policing strategically blends the features of multiple "operative logics"—prevention, deterrence, and preemption.[195] In one instance, predictive policing exists in "an objectively knowable world," where uncertainty is the product of data scarcity and the problem is simply acquiring enough data to maximize accuracy (predictive performance). At another moment, predictive policing leans heavily on certainty—the confidence that predictions lead to optimal distributions and resource allocations to deter crimes before they ever happen (prescription, or predictive performativity). And yet, in each operative configuration of prediction, uncertainty (indeterminacy) becomes such a pervasive, structural feature that the extent of its contagion into more complex analytic scaffoldings can never be fully known.

Most optimistically, we could read that indeterminacy as an opportunity to intervene in the patrol's ritualized enactment of state power. This was the HunchLab team's gambit. In a sort of double negation, the unknown-unknowns of crime occasion a chance to disrupt discriminatory patrol practices and geographies: indeterminacy divorces the patrol from its routinized patterns, what Rahman called "maintaining the status quo." In this sense, the crime predictions act as what media theorist Wendy Hui Kyong Chun has called an "ambivalent pharmakon"[196]—an opportunity to shuffle policing's future geographies through a data apparatus that challenges our assumptions of what constitutes evidence. Prediction may nominally be about representing and anticipating future events, but it works by capturing distributions of the present; and like forecasts of global warming's effects, we could use those indeterminant predictions as an opportunity to intercede, to untether the status quo of today from the ground truth of tomorrow.

In the end, however, this optimism is overwhelmed by the persistence of police discretion—the localized sovereignty of the "bobby," the constable, the patrol officer. Most critiques of predictive policing, for instance, tend to ignore the fact that existing best practices, including Compstat, are themselves crude algorithms of patrol geographies.[197] These blunt instruments have been the technique of choice for over-policing poor and minority communities, and the wide geographic berth of a strategy like "hot spot policing" leaves plenty of room for officer discretion. The HunchLab team is not wrong to frame their project as a corrective to this and other policing practices that allow police to determine who or what constitutes a threat to the social order. As social theorists Dubber and Valverde insist, the police's preventive power has always entailed discretion and localized sovereignty.[198] HunchLab's project was to rationalize, systematize, and optimize the implementation of that sovereignty, to introduce accountability for the discretion, to shuffle the patrol's logistical circulations to generate a more equitable distribution of public safety benefits. But with indeterminacy so thoroughly baked into the analytics, it became impossible to evaluate those redistributions without falling back on unfalsifiable claims or impossible trade-offs.

In his study of the Chicago Police Department's predictive policing efforts, critical geographer Brian Jordan Jefferson argues that the structure of the crime prediction apparatus "ensures that negatively racialized fractions of surplus labor and the places they inhabit are *only representable to state authorities and the public as objects of policing and punishment.*"[199] The HunchLab team may have tried their hardest to expand this narrow legibility, but ultimately, the product is a reform: it poses no fundamental challenge to the police patrol as a spatial economy of safety and danger. Every fiber of the social, technical, and institutional assemblage of policing is organized around "positive" outcomes—observable, quantifiable, optimizable. Letting-alone, letting-be, respecting—these are statistically inscrutable within the data structures and institutional formats available. Is it possible to reverse this illegibility and truly disrupt entrenched police abuses and violence? What would it look like to view HunchLab's interventionary trade-offs—between fairness and accuracy, exploration and exploitation—with an eye toward remediating disinvestment in Black and brown communities, as Siegel suggests? Imagining such a scenario is difficult because the ambiguities, indeterminacies, and trade-offs that plague predictive policing are *innate to the police patrol itself* as a logistical–ritual enactment of state power—not the effects of an algorithm.[200]

In October 2018, Azavea sold HunchLab to ShotSpotter, a police tech firm specializing in gunshot detection. Cheetham, Azavea's founder and CEO, authored a blog post explaining that the decision to sell was an ethical one.[201] Throughout 2017, HunchLab had been ramping up implementation with their newest and largest client, the Chicago Police Department. This was a drain on the programming team, with HunchLab demanding more time and effort than it was allotted. Proceeding would require taking on investors or private funders, and this would threaten Azavea's commitments to social betterment and transparency that had earned the company its B Corp certification. But Cheetham also explained that he could no longer justify the incongruities between those commitments and the larger project of predictive policing:

> Over the past several years, an array of incidents have publicly documented violence, civil rights violations, and abuse of power by police officers across the United States. Law enforcement agencies have rightfully come under increasing scrutiny. Further, "predictive policing" tools (license plate readers, facial recognition software, etc.) have been used in some communities

> to engage in pervasive surveillance of citizens, something that I believe is wrong. We're a B Corporation with a mission to not only advance the state of the art but also apply it for positive civic, social, and environmental impact. Developing a tool that would support surveillance or violate civil rights was not something I viewed as aligned with our mission.[202]

What Cheetham sensed here—and what ultimately led him to abandon a project that he oversaw for the better part of a decade—is not new. It is the same set of tensions that have plagued liberal-consent policing since the earliest commissioned police forces—the "delicate balancing of respectful protection and intrusive penetration." New police technologies emerge in attempts to resolve this tension, to tilt the balance back toward the respectful protection through better management, but the risk is that those same technologies will also occasion "a fuller and more penetrating gaze" into the polity.[203]

Ultimately, even HunchLab, the industry's greatest champion of predictive policing for reform, was incapable of resolving the patrol's contradictory logistical functions. On the one hand is the mandate to distribute public safety benefits as a collective good—an idea that traces back to early theorists of the modern state.[204] On the other is the view from marginalized communities, who, for generations, have experienced modern police institutions as an occupying force. On the first view, patrols act to safeguard the public from criminality; accordingly, a more equitable distribution of protections can be optimized for. In the latter view, whole communities and neighborhoods are criminalized (or at least rendered "criminogenic") through location-based patrols. The police are not protecting and serving—they are concentrating the law's uneven gaze in neighborhoods isolated by unjust policies, disinvestment, and renewal programs. For these communities, the logistical "optimization" of policing means just a more "effective" distribution of harm. That these two functions cannot be reconciled is not the fault of the algorithm per se. But neither can the algorithm offer any sort of meaningful resolution. Improving public safety benefits for all communities—enacting more equitable geographies of risk and protection—will require grappling with, reorganizing, even dismantling the entire sociotechnical and institutional apparatus of urban policing.

Conclusion

Toward an Urban Counterlogistics

> Logistics is more than the extension of the world market in space and the acceleration of commodital flows: it is the active power to coordinate and choreograph, the power to conjoin and split flows; to speed up and slow down.
>
> —JASPER BERNES, "Logistics, Counterlogistics and the Communist Prospect"

A Techno-Urban Supplement

The smart city is here—only, it does not look how you imagined it. According to a report from Big Four consulting firm Deloitte, smart cities are those that "organize around digital power" to "move from a linear to an exponential growth trajectory"—a sort of Moore's law for urban governance.[1] The McKinsey Global Institute likewise sees urban salvation in technology: as "millions of individual actors use data to make better decisions for themselves, the effects add up, causing the city as a whole to become more productive and responsive."[2] Such claims reveal the consulting class's ambitions for our urban futures, but they offer little insight into what life is actually like in cities today. They are reveries of a capitalist sublime that imagines cities as a deregulatory paradise, with entire neighborhoods, communities, and regions serving as test beds for the commodification of urban innovation's raw material—efficiency.[3] For that is what consultants do: they advise corporate managers, and now city leaders, on how to streamline.[4] They coordinate, plan, and identify new sources of value to be eked out of operations.[5]

When it comes to cities, though, those consulting dreams are nothing more than a fantasy, and the lengths to which city managers and industry leaders will go to realize such ambitions not only fall short of the mark—they have consequences for our communities. Whether we are talking about public spaces, service platforms, police patrols, or any of the countless infrastructures in the crosshairs of digital upgrading, the frictions, indeterminacies, and cultural politics of technological intervention reveal the smart city for what it's always been: a *fetishization* of precision, control, and logistical efficiency. Like commodity fetishism in Marx's analysis of capitalism, the smart city's obsession with efficiency mistakes the rich and uneven relationships of urban life for technical configurations—correlations between variables, optimizable effects, rationalized circulations, and so on. This masks the extent to which efficiency is itself a social relation, a *qualification* of human–machinic interactions "capable of producing desired results with little or no waste"[6]—where the determinations of what counts as "desired results" and "waste" can always be defined to suit the interests of powerful actors. And like commodities, efficiency's abstractions can become so dislocated, so disembedded from their social and material grounding in urban worlds, that the underlying relationships disappear from view entirely. Is it possible to rehabilitate efficiency's grounding in the social, political, and material ecologies of the city? To bring it down from the clouds of remote servers and consultants' dreamworlds to the street?

The city is a social plane, sociologist AbdouMaliq Simone writes, made up of "provisionally stitched together, jigged up intersections of bodies and materials upon which things are both moved and caught—a textured surface that speeds things up and slows them down."[7] If logistics designates the power to speed up and slow down, as poet Jasper Bernes suggests in the epigraph, then "urban problems" are those "textures" that undermine efficiency, that slow things down when they ought to be sped up, and vice versa. New technology is alluring as a corrective to this urban problematic. It appeals to an enduring view of the city as a space of both immanent virtue and immanent vice: in one instance, urban circulations represent a font of ingenuity, creativity, and productivity, but in another they pose a threat to morality, sociability, and stability.[8] Urban *governance* is the project of tilting the balance away from the vice, toward the virtuous; urban *politics* is the struggle to define the distinction between the two. The technologies of urban governance promise to wipe clean the residues from the city-as-surface, to smooth out the vicious frictions in the name of efficiency.[9]

But the smart city's technological "solutions" are always a *pharmakon*—both remedy and poison. They leave an excess that thrusts efficiency back into the realm of urban politics. Following philosopher Jacques Derrida, theorists Ryan Bishop and John W. P. Phillips describe this excess as the *urban supplement*—"those apparently inessential or marginal elements of an ensemble (like a city) that figure more necessary conditions, qualities without which the ensemble could not have arisen, but which often appear not only marginal but also threatening in various ways."[10] The smart city generates supplemental elements that at once motivate and confound efficiency interventions. Technologies embody their own dialectical double movement.[11] New frictions multiply as technicians and systems operators work to smooth out the surface, to rid the metropolis of efficiency's Others—idleness, neglect, irrationality, error, risk.

My theoretical gambit has been to foreground the techno-urban supplement, to examine the frictive elements of excess left behind by the smart city's commands of coordination, rationalization, optimization. Because when we privilege the supplement, the smart city's calculative mobilizations come into view as *contradictory enactments of political, economic, and social power,* always producing and negating the seeds of their own undoing. New efficiencies create new "inefficiencies"; new connections leave others disconnected; rationalizations produce new "irrationalities." LinkNYC's wireless connectivity may have privileged access and use, but the politics of the network's design generated categories of misuse and abuse that mapped onto existing exclusions by race and class; on-demand service platforms foretold greater autonomy and flexibility for workers but in practice enabled new forms of control through regulatory arbitrage, price manipulations, and information asymmetries; prediction promised greater precision in police resource management but wound up amplifying epistemological anxieties that emanate from the institution of policing itself.

These contradictory outcomes are not ancillary byproducts of miscarried calculations or mobilizations. They are the constitutive excess to the "smartness" of logistical governance—the "relatively free or unbound points, points of creativity, change and resistance" that both motivate and confound the smart city's logistical commands.[12] When we recognize logistics as (urban) power, the supplement demands our attention.[13]

Doing Logistical Governance

In the book's introduction, I argued that understanding the smart city as it "actually exists" requires accounting for the particularities of its awkward integrations while attending to the logics spanning its discrete interventions. The common denominator was logistics—calculative mobilizations that reconfigure urban life as an infrastructure for maximal efficiency, for the monetization of time and space.[14] But by what techniques does logistical governance manifest the smart city's apparent destiny by colonizing "the last great frontier of inefficiency in capitalism"?[15] What logics and strategies usher technology's awkward integration into the social order of the city—and how do technology producers deal with the supplement that threatens that integration? What, if I may, are the *logistics* of logistical governance?

We can discern four techniques that cut across the case studies' specificity. First, technological interventions require *legitimation.* In each chapter, we observed elite actors—city officials, politicians, pundits, professional designers, corporate reps, product managers, economists, police intellectuals, and so on—lay claim to social good or public benefit as justification for investment in new technologies or infrastructures. Indeed, it is rare to see appeals to "efficiency" alone as a sufficient rationale for resource expenditure. Even in McKinsey's pro-business formulation, efficiency enhancements are only ever as good as the ends they serve: "When cities function more efficiently, they also become more productive places to do business."[16] Key performance indicators, benchmarks, and "smartness quotients"—this business-speak is too cold, too instrumental, too calculating to garner sufficient goodwill among the necessary constituencies and publics.[17] Smart city boosters therefore try to put a human face on technologies, to craft more compelling narratives, to bind calculative mobilizations to moral and ethical ideals. The tokenistic participation of the Reinvent Payphones Challenge, for instance, created a sense of inclusivity and access around the project despite the relatively foregone conclusion of the technology's design; later, politicians would frame LinkNYC's connectivity as a way to combat economic inequality. In on-demand economics, flexibility took on an aura of morality as a respite to centuries of undemocratic employer control while papering over workers' on-the-ground experience of manipulation, misinformation, and control. And with predictive policing, we saw the HunchLab product team bind its intervention to police reform—indeed, the entire project of predictive policing grew out of a crisis of legitimacy in the institution of American policing. In each case, technology producers or promoters must establish a moral or ethical mandate to make their intervention palatable.

Second, technological interventions live and die by the social hierarchies they mediate, and these hierarchies map onto *asymmetrical regimes of information flow*. The ontologies of "system manager" and "end user" gloss subjects ranked by access. Like most proprietary information systems, such disparities are a defining feature of urban technologies. Each of the interventions examined in the book was black-boxed in one way or another, and this opacity generates asymmetries that are necessary for the technology's basic operability. Across the case studies, the information architecture had to remain concealed to end users.[18] Audiences, workers, and police officers may know only what they need to, only see what they are meant to—nothing more and nothing less. Such concealment cultivates what media sociologist David Hill describes as cognitive, social, and moral injuries.[19] Obscuring the information architecture means obscuring the material, infrastructural, and labor conditions necessary to support calculative mobilizations. Police are more likely to buy into an algorithm when they do not know that recommendations are based on randomized outputs; customers are more likely to order Caviar food delivery when they cannot see the labor and manipulation involved; New Yorkers are not in open rebellion because LinkNYC's data collection remains covert (despite journalists' best efforts).[20] The concealment of these infrastructural details, Hill argues, leads to "a pollution of the sensible" that attacks the very possibility of politics. It supports the fiction that "efficiency" is an objective condition, not a set of contestable social and material relationships.

Third, if the information asymmetries of black-boxed technologies mediate social hierarchies, then *modulation* is the technique by which those social distances are operationalized to achieve desirable effects. Modulation is the practical know-how of logistics, the statistical physics of efficiency—and what philosopher Gilles Deleuze's diagnoses as the quintessential operation of control. Unlike the rigid optics and confining architectures of discipline (epitomized in the Panopticon),[21] control's modulations configure an "open environment" (epitomized in the smart city)[22] through matrices and metrics—like "a sieve whose mesh will transmute from point to point."[23] And whereas disciplinary power engages the individual as subject, the smart city's controls operate on the *dividual,* the data point, the computable proxy.[24] For example, imagining a city where access is mediated by electronic card, Deleuze suggests that "what counts is not the barrier"—the gate, the entryway, the securitized lobby—"but the computer that tracks each person's position—licit or illicit—and effects a universal modulation." In the open environments of the city, those matrices and metrics create the conditions for optimization through modular concealments and revelations—ignorance and information—that allow system managers to observe, adjust, and refine effects under the guise of non-coercive "choice."[25]

We see these modular techniques applied to the smart city's dividual inhabitants in the various nudges, control levers, and customizable fields that system operators command to direct attention and movement, to persuade decisions without forcing them—when LinkNYC infers and acts on demographic information to create an "event"; when Caviar manipulates pay rates to incentivize workers to log on; when HunchLab directs patrol officers to areas they would not have thought to. Inputs and outcomes are carefully monitored, while the sieve of dividuation slices and dices statistical traits to realize effects on the aggregate level.

But once parsed by matrices and metrics, city dwellers' dividualized proxies and data doubles are only ever partial representations, impersonal and incomplete.[26] This *proximation* (making proxy) produces a supplement that will never be fully subsumed within the system—affective, qualitative intensities of desire, frustration, even boredom, that cannot be reduced by quantification to the digitality of urban data.[27] On the one hand, system administrators and infrastructure managers care little about this gap—the differential between data and the lifworlds they are meant to represent. What matters are "actionable" outcomes; whether the statistical proxies correspond to self-identifications or institutional categories of personhood is of no concern.[28] Effects trump fidelity. The "effect is the content," as Jeremy Packer puts it—transactions initiated, circulations optimized, routes rationalized, resources streamlined, productivity upped.[29] On the other hand, the fetish of efficiency requires work to sustain the fiction of its objectivity. Labor is needed to ensure that good-enough proxies get the job done, that they produce effects—especially, as we saw across the chapters, when the correspondence between the machine's output and "ground truth" data becomes dubious (as in the disinformation campaign of "demand throttling," predictive policing's indeterminate predictions, or the bizarre consumer-profiling inferences made possible on a system like LinkNYC).

Faith in the proxy is thus requisite, as the possibility for negotiation and nuance diminishes. The smart city propagates what sociologist Michalis Lianos and the late anthropologist Mary Douglas called "automated sociotechnical environments"—code/spaces where access, opportunity, and interaction are mediated by machine-readable tokens of authenticity and status.[30] As logistical governance expands this automated gatekeeping into new registers of urban life, institutional contact becomes reduced to "non-negotiative interactions." Limits are "set in advance and the whole existence of the user is condensed into specific legitimizing signals which are the only meaningful elements for the system."[31] If these systems could operate perfectly—if the fidelity of the proxy to its object, the dividual to the individual, were 100 percent accurate 100 percent of the time—we would have no need for social, cultural, political, or perhaps even economic capital: identity and identification would speak for themselves, and resources, opportunity, and access would be distributed with maximal efficiency. But this is precisely the smart city's fantasy. Technological systems *rarely* work as designed. As Deleuze surmised, the machine-readable card that grants access "could just as easily be rejected on a given day or between certain hours."[32] In instances of breakdown and misrecognition, the supplement rears its head; the social, the material, the political and cultural—all that seemed purged by the automated environment—return with a vengeance.

Which groups will have the latitude to exercise nuance, to negotiate their access, and who will be reduced to the machine-readable signals of their dividuated identity? The hierarchies of information access create the conditions for novel forms of *sociotechnical sovereignty*.[33] This is the fourth and final technique of logistical governance—the authority to decide on the exception,[34] to determine the privileged few who may speak for themselves while commanding that the rest be spoken for by their proxies. Such is the mundane sovereignty enacted in the everyday spaces and times of the city. And the discretion is scalar—from the individual patrol officer checking maps of crime risk in her squad car to the platform "deactivating" a worker because his performance falls below an arbitrary threshold, to the mayor's office advising their corporate partners to prevent information "abuse" by restricting functionality. Each exception signals an attempt to expunge the frictions—the supplemental—from the city's surfaces, to manage the excesses, the constitutive "outside" to logistical governance's efficiency fetish.[35]

Radical Logistics

If logistics describes the "active power to coordinate and choreograph," then what we see across the case studies are attempts to concentrate and monopolize that power—with society's least advantaged bearing the brunt of the cost and risk.[36] The question now is, What do we do with that knowledge? How do we translate this understanding of the techno-urban supplement into a radical urban politics? What role might efficiency and optimization play in efforts toward repair and maintenance?

Critical logistics thinkers like Bernes and theorist Alberto Toscano argue that the most potent resistance to global supply chain capitalism involves a "counter-logistics" of interruption and antagonism, a "logistics against logistics" that identifies and targets choke points in circulation regimes.[37] In the smart city, these choke points are both everywhere and nowhere. They do not concentrate geographically in the same way that the logistics industries concentrate in port cities like Dubai, Shenzhen, or even Oakland; they are embedded throughout the urban landscape, and this makes them both easier and more difficult to target for disruption. As the activist group Rethink LinkNYC has demonstrated, for example, it is possible to call out covert surveillance with something as simple as flyers duct-taped to street furniture. Similarly, as Marco Briziarelli shows in his account of a Deliveroo courier strike in Milan, platform workers have a number of tactics that they may use to frustrate the normal flow of information and money, such as logging on but rejecting orders en masse.[38] In the case of predictive policing, journalists, activists, and civil rights groups have been forced to act within the arcane informational space of public documents and city contracts, filing Freedom of Information Act requests (and bringing suit if necessary) to expose algorithms that in some cases remain secret even to city officials.[39]

These actions highlight potent sites and tactics of resistance to the smart city's encroachments. But in the end, they remain reactive. Their ingenuity is to use logistics against itself, against the material configurations of logistical power, but they do not coordinate new formations. What would an *active* strategy of counter-logistical power look like? This is the question that I would like to pose as a conclusion—not how city dwellers might protect themselves from technological harms, but how we might exploit the same "active power to coordinate and choreograph" to build a better city. Rather than subordinating the future to data—the measurable evidence of today's status quo—we need programs capable of *subordinating data to the future city*, and that do so in ways that upend assumptions about who or what designates the techno-urban supplement.[40] In part, the challenge is one of imagination. Because city leaders have been so thoroughly sold on consultants' fantasies, we are ill equipped to imagine urban technologies as anything other than efficiency amplifiers and optimization hacks. As a final remark, then, I would like to showcase two exemplars for how we might reimagine urban technologies as a means of support and repair, a way to make life more livable for society's most vulnerable communities.

Sanctuary in the Streets: A Counter-Logistics of Community

The post-2016 rise of authoritarian strongmen governments cast the smart city control apparatus in a new and more threatening light. Donald Trump built his 2016 presidential campaign on a nativist, anti-immigrant, and "law and order" platform, routinely praising police officers' abuses; in office, his policies and appointments reflect commitments not to workers (as he pledged on the campaign trail) but to corporate executives and finance capitalists, whom he hails as job creators. Most terrifying are Trump's dogged efforts to target and deport immigrants fleeing violence and poverty to seek asylum in the United States. Under Trump, the Immigrations and Customs Enforcement (ICE) agency has initiated a crackdown on undocumented migrants living in the United States, and especially those in "sanctuary cities"—where local and state government institutions refuse to comply with ICE directives for information or prisoner transfer. Trump has authorized ICE field agents a much greater degree of discretion than prior administrations, leading to surging rates of arrests among noncriminal migrants, breaking up families, putting children in cages, and leaving federal courts clogged.[41] While average deportations per month are actually down relative to the Obama administration, private prisons are filling up with migrants awaiting sentencing or deportation.[42] Indeed, stocks in the corporations running these vast prison complexes have risen dramatically, with government contracts and new facilities opening to manage the overflow of immigrant detainees.[43]

In 2014, ICE contracted with secretive Silicon Valley smart city analytics firm Palantir to update and expand its Investigative Case Management (ICM) system.[44] Cofounded by Peter Thiel, a transition adviser to Donald Trump, and with investment from In-Q-Tel, the venture capital arm of the Central Intelligence Agency, Palantir has enjoyed numerous government contracts since its launch in 2004—from predicting the locations of improved explosive devices (IEDs) in Fallujah, to stock market analysis and predictive policing in Los Angeles and New Orleans.[45] In contrast to the Treasury Enforcement Communication System—the data management system that ICE and its predecessor, the Immigration and Naturalization Service, had used since 1987—Palantir's new ICM system for ICE integrates multiple federal law enforcement databases, including the Drug Enforcement Agency; the Bureau of Alcohol, Tobacco, Firearms, and Explosives; and the Federal Bureau of Investigation.[46]

This ICM is subject to severe mission creep, with its original purview expanding from Homeland Security Investigations (the investigative arm of the Department of Homeland Security) to ICE's Enforcement and Removal Office (ERO). It represents the kind of "data fusion" that federal agencies have been promoting since the 9/11 attacks.[47] In practice, such fusion will enable ICE–ERO agents to track a suspect across state, local, and federal government databases, including anything from biometric information (fingerprints, retinal scans, tattoos) to automobile registration and border-crossing histories, streamlining the process for obtaining search, seizure, and arrest warrants. With Palantir's help, ICE is becoming nothing short of an efficient machine for mass deportations.[48]

In the face of this authoritarian logistical regime, an initiative called Sanctuary in the Streets shows what a grassroots counter-logistical infrastructure might look like. Sanctuary in the Streets enlists community members not at risk of deportation to engage in protest and other civilly disobedient actions to disrupt ICE raids in progress. In Philadelphia, the Sanctuary in the Streets program is coordinated by the New Sanctuary Movement, an interfaith social justice organization providing shelter, legal aid, and community support to immigrants, with sister operations in western Massachusetts and elsewhere. The effort dates to the Obama administration and its handling of the 2014 migrant surge—it is easy to forget in the post-2016 era that when the U.S. southern border received waves of children fleeing violence in Central American countries, it was Obama who ramped up deportation efforts, well before Trump was ever in office.[49] Nevertheless, with broad dismay at Trump's campaign rhetoric, Sanctuary in the Streets gained momentum in the months immediately following the 2016 election; by early 2017, the number of active members in the Sanctuary in the Streets network jumped from around sixty to more than a thousand.[50]

Real-time protest coordination to disrupt ICE raids in progress is not easy. ICE is incredibly secretive of its agents' movements and raid plans. Unlike municipal police departments, they do not operate on radio frequencies open to the public. How, then, to organize a crowd sizable enough and rapidly enough to disrupt a raid or arrest in progress? The New Sanctuary Movement distributes hotline numbers to members of local immigrant communities through immigrant-serving institutions (such as churches and places of work); a volunteer network-member staffs a designated cell phone, fielding calls twenty-four hours a day. When ICE initiates an arrest or raid, witnesses or family members can report it through the hotline; the volunteer manning the phone then sends an SMS and email alert throughout the activist network, relaying information about the time and location of the arrest or raid. In many cases, the arrests take place during scheduled check-ins with ICE agents or immigration-court representatives,[51] so a number of the actions have centered on ICE's Philadelphia field offices. Once the information is out, network-activists converge on the site to take part in song and prayer; meanwhile, a smaller group of individuals who have committed to risking arrest engages in carefully planned tactics designed to physically disrupt the arrest's progress, including sit-downs and the formation of human chains.

Bad Batch Alert: A Logistics of Harm Reduction

Trump's inhumane scapegoating of immigrants penetrates nearly every one of his administration's policy initiatives. In March 2018, Trump gave a forty-minute speech titled "Opioids: The Crisis Next Door." The United States had experienced a nearly 10 percent increase in opioid-related deaths over the previous year: more than seventy thousand Americans overdosed on opiates, including heroin, prescription painkillers, and variations on synthetic cannabinoids and the synthetic opioid fentanyl, which is often laced into batches of heroin.[52] Unlike the moral panic around crack cocaine that ramped up the War on Drugs in the 1980s and which led to the mass incarceration of African Americans as well as members of other racial-minority communities,[53] the "opioid crisis" largely affects poor white Americans, many of whom are Trump supporters that live and vote in Republican strongholds. With these racial politics, Trump was hardly about to indict his own base, and instead drew a direct line to his tired reproach of immigrants and their communities. "Ending sanctuary cities," Trump exclaimed in the speech, "is crucial to stopping the drug addiction crisis."[54]

The absurdity of this statement cannot be overstated: it lets the forces actually driving the epidemic—the billion-dollar pharmaceutical firms and insurance agencies, the pill mills and overprescribing doctors, and so on—off the hook while falsely laying the blame on urban policies and the immigrant communities they are designed to protect. And contrary to Trump's asinine claims, cities have been at the avant-garde of the fight against opioid-related deaths. While state attorneys general have been mounting lawsuits against pharmaceutical companies for incentivizing doctors to overprescribe painkillers,[55] city health departments have been working on the ground to develop harm-reduction programs, many already proven successful at preventing unnecessary overdose deaths. Harm reduction describes a set of principles and practices to staunch the "real and tragic harm and danger" associated with both licit and illicit drug use.[56] It avoids moralizing drug use by acknowledging a continuum from severe abuse to total abstinence, without imposing a binary on those two extremes. The goal, as the name suggests, is to reduce the harm that drugs may cause to users, their friends and family, by "meeting people where they're at," without casting judgment.[57]

One example: librarians in Philadelphia and other cities have requested trainings on the administration of naloxone, a life-saving opioid overdose reversal spray.[58] As "drug tourists" from across the country funnel into Philadelphia's Kensington neighborhood, where the open-air drug market (known as the "Walmart of heroin") is famous for its pure, high-quality heroin, librarians at the local McPherson branch have been on the front lines as a life-saving infrastructure of support and care.[59] The McPherson librarians organized trainings with volunteers and activists at the nonprofit Prevention Point Philadelphia, the city's longest-running needle exchange and distribution program, and library staff from branches across the city showed up. The trainings linked the library's public spaces and resources with Prevention Point's decades of experience in harm reduction and direct action. The staff at McPherson now runs "overdose drills," ensuring that everyone on hand knows the most effective course of action to save a life during an opiate overdose. The model's success is evident in the growing number of libraries across the country adopting similar approaches.[60]

Of course, not all overdoses happen in or near a library, and librarians alone cannot carry the weight of fighting the opioid epidemic. City health officials and epidemiologists track overdoses and leverage that data to identify geographic clusters of "bad batches," where heroin that is more potent than usual or laced with harmful additives such as fentanyl or synthetic cannabinoids (often referred to as K2) turns up in a pattern. Indeed, this was the case in September 2018, when more than two hundred individuals in Kensington were exposed to a potentially toxic combination of heroin and K2. The city had just begun trialing a new Opioid Overdose Notification system, which allowed officials to push texts and email alerts to first responders and members of the harm-reduction community to help them identify bad batches by their unique ensemble of symptoms and by the individual sources thought to be responsible for the majority of overdoses.[61]

Philadelphia's overdose notification program was not the first of its kind. It was inspired a similar system used in Baltimore since August 2017 called Bad Batch Alert. Developed by six high school students enrolled in Code in the Schools, a program that trains teens on coding and web development, Bad Batch Alert sources information from and directs alerts to notify communities when tainted heroin is circulating in their area. So whereas the Philadelphia program sends notifications to health officials and frontline workers in the harm-reduction community (including the librarians), Bad Batch Alert communicates directly with subscribers, and anyone can subscribe—users themselves or their friends and family. After epidemiologists with Baltimore's Behavioral Health Systems identify an overdose spike or cluster of incidents, the system triggers a location-specific alert. Anonymous subscribers signed up by text or email then receive a notification that a bad batch of heroin has entered the neighborhood.[62] The system includes a number of built-in commands that allow users to access services and support in their area, including connection to a twenty-four-hour crisis line, real-time location services for a needle exchange and distribution van, and information about upcoming naloxone trainings. "With this knowledge," the program's teenaged inventors suggest, "dosages can be modified and loved ones can be warned," enabling those most immediately affected by the tainted batches to better protect themselves.

Urban Technologies of Care

Systems like Bad Batch Alert and Sanctuary in the Streets represent another smart city, one that leverages technology to do the difficult work of repair and maintenance with communities too often cast as "inessential or marginal elements." But these are exactly the types of programs that get left out of much smart city discourse. For example, in 2017, the City of Philadelphia launched the SmartCityPHL initiative. Funded in part by the John S. and James L. Knight Foundation and produced in consultation with Big Four consulting firm Guidehouse LLP (formerly PricewaterhouseCoopers), the initiative convened a working group to define best practices and determine a course of action to realize Philadelphia's smart city potential. Representatives came from across city government—the Department of Streets, the Office of Sustainability, the Fire Department, the Departments of Public Property and Licenses and Inspections, and so on—culminating with the 2019 publication of a roadmap, "an initial guide to spur innovation and collaboration in City government around smart city [*sic*] and the policies and technology surrounding it."[63]

SmartCityPHL presents a future in which technology integrates and enhances the city's existing digital assets, including gunshot-detection systems, police-worn body cameras, traffic cams, automatic vehicle locations systems, lighting controls, metering, health image processing, trash-can and stormwater sensors, transit data, and more. Such integrations are deemed worthwhile in their own right, but according to the roadmap can become even more valuable "as more of the city's built environment becomes tech-enabled and requires the support of a network."[64] As the city's Assistant Director for Strategic Initiatives Ellen Hwang put it, "It isn't just about mobility, it isn't just about public health—it's like, 'how do all these specific topic areas within smart cities tie together and how do we as an IT department better support the foundation to build all those different initiatives?'"[65]

The goal is not trivial and the program's intentions are certainly laudable, but who, exactly, are these initiatives meant to serve? Several communities were notably absent from the roadmap's plans. The only mention of Philadelphia's large and still-growing immigrant communities—about 15 percent of the city's entire population, 19 percent of its workforce, and 14 percent of all Philadelphians living in poverty[66]—comes in a brief discussion of a language-accessibility pilot program that would use "voice-powered technology," with data "collected, mined, and analyzed to capture opinions, ideas, and thoughts from the community."[67]

In this and other instances, the roadmap clearly succeeds in parroting the Silicon Valley business-speak of data monetization (with terms like "mined" and "capture" evoking an extractivist sensibility), but at no point are the city's opioid crisis, high poverty rates, or chronic homelessness mentioned. And while it may not be entirely fair to chide the SmartCityPHL initiative for these deficiencies (after all, I have just spent an entire book critiquing urban technological solutionism), they nonetheless exemplify the narrowness of officials' vision for Philadelphia as a smart city. Sanctuary in the Streets and Bad Batch Alert represent an alternative, more inclusive path. They show that technology can be used to coordinate, organize, and—indeed—optimize the allocation and circulation of information, but that coordination and allocations can be done in ways that support community action and care. Sanctuary in the Streets mobilizes protest as a last-stop measure to prevent the arrest and deportation of noncriminal community members, family members, and workers. Bad Batch Alert circulates critical information to prevent overdose deaths among communities struggling with often-debilitating addiction. They are simple systems, but they are powerful. They enact a critical logistics of information, the circulation of which has the potential to save lives. And it is clear that our inability to conceive of them as "smart city" technologies reflects a profound failure of our collective imagination and political will, a cultivated resignation to turnkey, proprietary systems that require compromises in privacy, autonomy, transparency, and accountability. Neither Bad Batch Alert nor Sanctuary in the Streets is likely to generate much revenue and therefore will not land any financing from Wall Street or the venture capitalists of Sand Hill Road; neither translates efficiency and calculative mobilizations into profit, nor do they promise politically palatable technical fixes in lieu of meaningful institutional reforms. And yet both commandeer information and communication toward vital ends; both define efficiency in terms of moral responsibility and an ethics of care, repair, and support.

As critical technology scholar Steven Jackson argues, one of the primary ideological dogmas driving the story of technology is a "productivist" view of innovation, where repair, support, and maintenance are secondary functions at best. Like the cognitive, social, and moral injuries of infrastructural concealment discussed above, the productivist view makes us blind to all the labor that goes into making technologies work, not only for efficiency's profits, but for people. Jackson writes:

> The language of innovation is generally reserved for new and computationally intensive "bright and shiny tools," while repair tends to disappear altogether, or at best is relegated to the mostly neglected story of people (researchers, information managers, beleaguered field technicians)

> working to fit such artifacts to the sticky realities of field-level practices and needs. In both cases, dominant productivist imaginings of technology locate innovation, with its unassailable standing, cultural cachet, and valorized economic value, at the top of some change or process, while repair lies somewhere else: lower, later, or after innovation in process and worth.[68]

In part, Jackson's contention is that media and technology research must reorient from the "bright and shiny tools" of technological innovation and toward the mundane labor of repair (in the narrow sense) required to sustain technological lifeworlds. But we can extrapolate from his argument a call to think through the ways in which everyday social technologies, faded into the background of our consciousness (hotlines, apps, epidemiological reporting and mapping, existing networks of activists and community organizers, and, indeed, payphones), can be enrolled in much more ambitious logistical projects to *challenge the status quo.* As sociologists and planning scholars Sung-Yueh Perng and Sophia Maalsen find, urban intelligence—"smartness"—can and will be "reinterpreted, coopted, and appropriated as it materializes in the actually existing city."[69] Through diverse practices of retrofitting, repurposing, and reinvigorating, a humane smart city subordinates "innovation" to the civic infrastructures the technology supports. In the same spirit, networks like Bad Batch Alert and Sanctuary in the Streets show how existing systems can facilitate new circulations, new coordinations, not only to resist corporate-driven interventions, but to actively redirect civic power and informational resources to those supplemental quarters of the city that have for too long been excluded from logistics' benefits.

Logistics, remember, is an agnostic art. We must be vigilant about its application. Whose efficiency, what is being optimized, and toward what end? If we can start seeing systems like Sanctuary in the Streets and Bad Batch Alert as smart city technologies—as a logistics of care and repair, as last-ditch infrastructural supports for communities' survival—then smart urbanism might start to look vastly different, and far more compassionate.

Acknowledgments

I write in the Introduction that the book started in transit, but this is an understatement. The project was shaped and nurtured by an intellectual community of comrades and mentors who provided an infrastructure of support and shepherded me through an education. Without that community, not a single page could have been written. The book is an artifact of that community's influence. First, I thank my professional guides on this journey. As my research interests evolved throughout the PhD program at Penn's Annenberg School for Communication, they were always kept in check by the steady hand of my adviser, Carolyn Marvin. From Carolyn I learned what a good research question looks like, what good writing is (and how much work is involved), and, most important, how to anticipate challenges to my argument—in other words, the craft of the profession. Jessa Lingel is my other mentor (and comrade), and I am indebted to her close attention to—and generous support of—my research over the years. From Jessa I continue to learn how to be a caring citizen-scholar. I also thank my dissertation committee for their input and encouragement: John Jackson, who has advised and inspired me since my undergrad days as an anthropology major; Marwan Kraidy, who anticipated and challenged me to develop the "logistical" thread running through the research; and Mimi Sheller, who, as an external committee member, helped me tailor my work to an interdisciplinary audience and, specifically, gave helpful guidance on chapter 2 of the book.

I would be remiss if I did not also acknowledge the broader community of faculty, scholars, staff, and peers at the Annenberg School whose influence is visible perhaps only to me in the pages of this book. Many of my journeys, literal and intellectual, trace to the Scholars Program in Culture and Communication under the direction of Barbie Zelizer and the care of Emily Plowman. Not only did the program support unparalleled travel and research opportunities, it introduced me to visiting scholars whose supportive attention to my work opened up new ways of seeing the world, including Lisa Parks, John Nerone, and Kelly Gates, who helped tremendously with chapter 3. To this end, I must also acknowledge Michael Delli Carpini who, as Annenberg's dean, supported this and other programs (including the student-run CultureLab and 3620 Podcast) that provided a scholarly home for me and my peers on the "culture" side of things. In the same regard, I thank Guobin Yang and Sharrona Pearl for always having students' backs, and Sharrona in particular for helping me and others adjust to graduate school life. The Annenberg School's amazing staff are what make the institution a community. I thank in particular Sharon Black, Min Zhong, Joanne Murray, Kelly Fernandez, Rose Halligan, Margie Boylan, Litty Paxton, Deb Porter, and my teaching partner Waldo Aguirre. Finally, there is no way I could have survived my education without the support of my peers and friends, each of whom provided feedback over the years in one form or another: Omar Al-Ghazzi, Osman Balkan, Lyndsey Beutin, Lauren Bridges, Rosie Clark-Parsons, Zane Cooper, Jasmine Erdener, Nick Gilewicz, Ashley Gorham, Kevin Gotkin, Sun-Ha Hong, Natalie Herbert, Cameron Hu, Sanjay Jolly, David Karp, Emily LaDue, Corrina Laughlin, Josh Lewis, Tim Libert, Deb Lui, Lee McGuigan, Jasmine Salters, John Vilanova, Allie Volinsky, and Dror Walter.

During and after graduate school, I received financial and institutional support from a number of organizations, which gave me the physical and intellectual space to develop my research agenda and thinking. In particular, I thank Kate Crawford, Solon Barocas, Hannah Wallach, and the staff at Microsoft Research New York City (as well as Arvind Narayanan during his time there); Todd Wolfson, Victor Pickard, and Briar Smith at the Media, Inequality, and Change (MIC) Center at Penn and Rutgers; and Kathy Strandburg and Ira Rubinstein at the New York University Law

School's Information Law Institute (ILI), as well as my ILI peers Seb Benthall, Gabe Nicholas, Mark Verstraete, and Salome Viljoen. I am also thankful for my future colleagues at UNC who have taken a chance on me, especially Lawrence Grossberg, Alice Marwick, Steve May, Chris Lundberg, and Patricia Parker, as well as Maggie Franz.

Throughout the course of this book, we hear from designers, platform workers, and technology developers. I would like to thank all of the designers I interviewed, but especially Alex, Meredith, Nick, Ann, Alexei, and Will. The platform workers' names have been changed to ensure privacy and candidness. I thank them all for their helpful input, but especially Gary, Suzanne, and Mike. At Azavea, I thank Chip, Adele, and Robert, as well as the members of the geospatial programming team for their guidance; I would like to acknowledge Jeremy in particular for being so welcoming and patient with me as an ethnographer out of my element.

Of course, the book would quite literally not be in your hands without the support and encouragement of the University of Minnesota Press, and in particular the editorial team of Pieter Martin and Anne Carter, who were my guides in the process of bringing the manuscript to fruition; Nicholas Taylor, who helped me overcome some of my more confusing turns of phrase; and Shannon Mattern and Matt Wilson, who saw potential through the clutter of early drafts.

Finally, there are those who are not part of my academic life but who nonetheless provided the care and support needed to get this thing done: Babette, my ma, and my dad, Jon; my siblings, Gabi, Jess, and Ria; my Grandpa Jay and Grammy Marion; Dave, Bitsy, Jake, Will, and Maddie; and my other family at Joe Ranch—Lily, Richard, James, Felicity, Big John, Betsy, Ben, Scott, Josh, Charlotte, Matt, Alan, Nick, and Avery. But most of all, this book is dedicated to Natalie, who humors me when I complain and helps me find solutions when I am actually struggling. I thank her not only for coping with big changes in our lives but for her genuine excitement about the prospect. I thank her for being a supermom to Nico. But mostly I thank her for being my partner through everything.

Notes

Introduction

1. Nvidia, "Smart City Analytics and Applications from Nvidia Metropolis," 2019, https://www.nvidia.com/en-us/autonomous-machines/intelligent-video-analytics-platform/.

2. Nokia, "Integrated Operations Center for Smart Cities," 2019, https://www.nokia.com/networks/services/integrated-operations-center-for-smart-cities/.

3. Richard Florida, "The Rise of 'Urban Tech,'" *CityLab,* July 10, 2018, https://www.citylab.com/life/2018/07/the-rise-of-urban-tech/564653/.

4. Allied Market Research, "Smart Cities Market to Garner $2.402 Trillion by 2025 with CAGR of 21.28%, Globally: Allied Market Research," *GlobeNewswire,* March 4, 2019, http://www.globenewswire.com/news-release/2019/03/04/1746100/0/en/Smart-Cities-Market-to-Garner-2-402-Trillion-By-2025-With-CAGR-of-21-28-Globally-Allied-Market-Research.html.

5. Ayona Datta, "A 100 Smart Cities, a 100 Utopias," *Dialogues in Human Geography* 5, no. 1 (2015): 49–53.

6. Ellen P. Goodman and Julia Powles, "Urbanism under Google: Lessons from Sidewalk Toronto," *Fordham Law Review* 88 (2019): 457–98; Bianca Wylie, "Searching for the Smart City's Democratic Future," Centre for International Governance Innovation, August 13, 2018, https://www.cigionline.org/articles/searching-smart-citys-democratic-future.

7. Clare Lyster, *Learning from Logistics* (Basel: Birkhäuser, 2016).

8. Interview with software developer and computer programmer, Azavea's offices, October 30, 2015.

9. Brett Neilson, "Five Theses on Understanding Logistics as Power," *Distinktion: Journal of Social Theory* 13, no. 3 (2012): 323.

10. Robert Cowley and Federico Caprotti, "Smart City as Anti-Planning in the UK," *Environment & Planning D: Society & Space* 37, no. 3 (June 2019): 428–48; Shannon Mattern, "Methodolatry and the Art of Measure," *Places Journal*, November 2013, https://doi.org/10.22269/131105; Matthew Zook, "Crowd-Sourcing the Smart City: Using Big Geosocial Media Metrics in Urban Governance," *Big Data & Society* 4, no. 1 (2017), https://doi.org/10.1177/2053951717694384.

11. Sarah Holder, "What Cities, Renters, and Employees Lose When Airbnb Comes to Town," *CityLab*, February 1, 2019, https://www.citylab.com/equity/2019/02/study-airbnb-cities-rising-home-prices-tax/581590/; Charles Maldonado, "Affordable Housing Group Says Fears about Airbnb's Effect on New Orleans Housing Have Come True," *The Lens NOLA*, March 28, 2018, https://thelensnola.org/2018/03/28/affordable-housing-group-says-fears-about-airbnbs-effect-on-new-orleans-housing-have-come-true/; Emily Peck and Charles Maldonado, "How Airbnb Is Pushing Locals from New Orleans' Coolest Neighborhoods: 'We're Out of Here,'" *The Advocate*, November 4, 2017, https://www.theadvocate.com/new_orleans/news/article_e8e96c00-c0e9-11e7-b886-cf6fbb846f57.html.

12. New York Taxi Workers Alliance, "8th NYC Professional Driver Dead by Suicide," November 14, 2018, http://www.nytwa.org/statements/2018/11/14/november-14-2018-8th-nyc-professional-driver-dead-by-suicide; Emma G. Fitzsimmons, "Why Are Taxi Drivers in New York Killing Themselves?" *New York Times*, December 2, 2018, https://www.nytimes.com/2018/12/02/nyregion/taxi-drivers-suicide-nyc.html; Brian M. Rosenthal, "As Thousands of Taxi Drivers Were Trapped in Loans, Top Officials Counted the Money," *New York Times*, May 19, 2019, https://www.nytimes.com/2019/05/19/nyregion/taxi-medallions.html.

13. Rob Kitchin, "Making Sense of Smart Cities: Addressing Present Shortcomings," *Cambridge Journal of Regions, Economy & Society* 8, no. 1 (2015): 131–36; Rob Kitchin, "The Real-Time City? Big Data and Smart Urbanism," *GeoJournal* 79, no. 1 (2014): 1–14; Rob Kitchin, Tracey P. Lauriault, and Gavin McArdle, eds., *Data and the City* (New York, NY: Routledge, 2017); Simon Marvin, Andres Luque-Ayala, and Colin McFarlane, eds., *Smart Urbanism: Utopian Vision or False Dawn?* (New York: Routledge, 2015).

14. Robert G. Hollands, "Will the Real Smart City Please Stand Up?" *City* 12, no. 3 (2008): 303–20.

15. Jathan Sadowski and Roy Bendor, "Selling Smartness: Corporate Narratives and the Smart City as a Sociotechnical Imaginary," *Science, Technology, & Human Values* 44, no. 3 (2019): 541.

16. Ola Söderström, Till Paasche, and Francisco Klauser, "Smart Cities as Corporate Storytelling," *City* 18, no. 3 (2014): 307–20; Francisco Klauser, Till Paasche, and Ola Söderström, "Michel Foucault and the Smart City: Power Dynamics Inherent in Contemporary Governing through Code," *Environment & Planning D: Society & Space* 32, no. 5 (2014): 869–85; Alberto Vanolo, "Smartmentality: The Smart City as Disciplinary Strategy," *Urban Studies* 51, no. 5 (2014): 883–98.

17. Sadowski and Bendor, "Selling Smartness," 541. See also James Merricks White, "Anticipatory Logics of the Smart City's Global Imaginary," *Urban Geography* 37, no. 4 (2016): 572–89.

18. Adam Greenfield, *Against the Smart City* (Helsinki: Do Projects, 2013); Robert G. Hollands, "Critical Interventions into the Corporate Smart City," *Cambridge Journal of Regions, Economy & Society* 8, no. 1 (2015): 61–77; Shannon Mattern, "Instrumental City: The View from Hudson Yards, circa 2019," *Places Journal,* April 26, 2016, https://doi.org/10.22269/160426; Molly Sauter, "Google's Guinea-Pig City," *The Atlantic,* February 13, 2018, https://www.theatlantic.com/technology/archive/2018/02/googles-guinea-pig-city/552932/.

19. Richard Sennett, "No One Likes a City That's Too Smart," *The Guardian,* December 4, 2012, http://www.theguardian.com/commentisfree/2012/dec/04/smart-city-rio-songdo-masdar.

20. Orit Halpern, *Beautiful Data: A History of Vision and Reason since 1945* (Durham, N.C.: Duke University Press, 2014).

21. Orit Halpern et al., "Test-Bed Urbanism," *Public Culture* 25, no. 2 (2013): 290.

22. Goodman and Powles, "Urbanism under Google."

23. Taylor Shelton, Matthew Zook, and Alan Wiig, "The 'Actually Existing Smart City,'" *Cambridge Journal of Regions, Economy & Society* 8, no. 1 (2015): 13–25.

24. Bruce Sterling, "Stop Saying 'Smart Cities,'" *The Atlantic,* February 12, 2018, https://www.theatlantic.com/technology/archive/2018/02/stupid-cities/553052/.

25. Linda Poon, "Songdo, South Korea's Smartest City, Is Lonely," *CityLab,* June 22, 2018, https://www.citylab.com/life/2018/06/sleepy-in-songdo-koreas-smartest-city/561374/.

26. Leyland Cecco, "'Surveillance Capitalism': Critic Urges Toronto to Abandon Smart City Project," *The Guardian,* June 6, 2019, https://www.theguardian.com/cities/2019/jun/06/toronto-smart-city-google-project-privacy-concerns; Sidney Fussell, "The City of the Future Is a Data-Collection Machine," *The Atlantic,* November 21, 2018, https://www.theatlantic.com/technology/archive/2018/11/google-sidewalk-labs/575551/.

27. See also Cowley and Caprotti, "Smart City as Anti-Planning in the UK."

28. Kaleigh Rogers, Brian Anderson, and Jason Koebler, "Kansas City Was First to Embrace Google Fiber, Now Its Broadband Future Is 'TBD,'" *Vice,* August 30, 2017, https://www.vice.com/en_us/article/xwwmp3/kansas-city-was-first-to-embrace-google-fiber-now-its-broadband-future-is-tbd.

29. Shelton, Zook, and Wiig, "The 'Actually Existing Smart City.'"

30. Alan Wiig, "The Empty Rhetoric of the Smart City: From Digital Inclusion to Economic Promotion in Philadelphia," *Urban Geography* 37, no. 4 (May 2016): 535–53.

31. Shelton, Zook, and Wiig, "The 'Actually Existing Smart City,'" 14.

32. Carolyn Marvin, *When Old Technologies Were New: Thinking about Electric Communication in the Late Nineteenth Century* (Oxford, UK: Oxford University Press, 1988).

33. Anna Lowenhaupt Tsing, *Friction: An Ethnography of Global Connection* (Princeton, N.J.: Princeton University Press, 2005); Nicky Gregson, "Logistics at Work: Trucks, Containers and the Friction of Circulation in the UK," *Mobilities* 12, no. 3 (2017): 343–64; Nicky Gregson, Mike Crang, and Constantinos N Antonopoulos, "Holding Together Logistical Worlds: Friction, Seams and Circulation in the Emerging 'Global Warehouse,'" *Environment & Planning D: Society & Space* 35, no. 3 (2017): 381–98.

34. Kate Hepworth, "Enacting Logistical Geographies," *Environment & Planning D: Society & Space* 32, no. 6 (2014): 1129.

35. Alan Wiig, "IBM's Smart City as Techno-Utopian Policy Mobility," *City* 19, nos. 2–3 (2015): 258–73.

36. Suhail Malik, "Information and Knowledge," *Theory, Culture & Society* 22, no. 1 (2005): 29–49; Sadowski and Bendor, "Selling Smartness"; White, "Anticipatory Logics."

37. Deborah Cowen, *The Deadly Life of Logistics: Mapping Violence in Global Trade* (Minneapolis: University of Minnesota Press, 2014); David Morley, *Communications and Mobility: The Migrant, the Mobile Phone, and the Container Box* (Hoboken, N.J.: Wiley-Blackwell, 2017).

38. Jasper Bernes, "Logistics, Counterlogistics and the Communist Prospect," *Endnotes* 3 (2013), https://endnotes.org.uk/issues/3/en/jasper-bernes-logistics-counterlogistics-and-the-communist-prospect.

39. Anja Kanngieser, "Tracking and Tracing: Geographies of Logistical Governance and Labouring Bodies," *Environment & Planning D: Society & Space* 31, no. 4 (2013): 594–610; Agnieszka Leszczynski, "Speculative Futures: Cities, Data, and Governance beyond Smart Urbanism," *Environment & Planning A: Economy & Space* 48, no. 9 (2016): 1691–1708.

40. Florida, "Rise of 'Urban Tech.'"

41. Sandro Mezzadra and Brett Neilson, "On the Multiple Frontiers of Extraction: Excavating Contemporary Capitalism," *Cultural Studies* 31, nos. 2–3 (May 2017): 185–204.

42. Antoine Henri de Jomini, *The Art of War,* trans. G. H. Mendell and W. P. Craighill (Rockville, Md.: Arc Manor, 2007), 51.

43. Michel de Certeau, *The Practice of Everyday Life,* trans. Steven Randall (Berkeley: University of California Press, 1984). One exception to this is Paul Virilio, who began theorizing logistics as power in the late 1970s; see Paul Virilio, *Speed and Politics: An Essay on Dromology* (1977; repr., New York: Columbia University, 1986); see also Neilson, "Five Theses."

44. W. Bruce Allen, "The Logistics Revolution and Transportation," *Annals of the American Academy of Political and Social Science* 553 (1997): 106–16; Cowen,

Deadly Life of Logistics; Martin Danyluk, "Capital's Logistical Fix: Accumulation, Globalization, and the Survival of Capitalism," *Environment & Planning D: Society & Space*, 36, no. 4 (2018): 630–47; Edna Bonacich and Jake B. Wilson, *Getting the Goods: Ports, Labor, and the Logistics Revolution* (Ithaca, N.Y.: Cornell University Press, 2008).

45. James R. Beniger, *The Control Revolution: Technological and Economic Origins of the Information Society* (Cambridge, Mass.: Harvard University Press, 1986); Alfred D. Chandler, *The Visible Hand: The Managerial Revolution in American Business* (Cambridge, Mass.: Harvard University Press, 1977).

46. Allen, "Logistics Revolution and Transportation," 110.

47. Neilson, "Five Theses."

48. Anna Tsing, "Supply Chains and the Human Condition," *Rethinking Marxism* 21, no. 2 (2009): 148–76.

49. Charmaine Chua et al., "Introduction: Turbulent Circulation: Building a Critical Engagement with Logistics," *Environment & Planning D: Society & Space* 36, no. 4 (2018): 617–29.

50. Bernes, "Logistics, Counterlogistics and the Communist Prospect;" Lyster, *Learning from Logistics*; Marion Fourcade and Kieran Healy, "Seeing Like a Market," *Socio-Economic Review* 15, no. 1 (2017): 9–29.

51. Quoted in Paul Rabinow, *Anthropos Today: Reflections on Modern Equipment* (Princeton, N.J.: Princeton University Press, 2003), 18; Michel Foucault, *Les mots et les choses* (Paris: Editions Gallimard, 1966), 670.

52. Ned Rossiter, *Software, Infrastructure, Labor: A Media Theory of Logistical Nightmares* (New York: Routledge, 2016), xvii, emphasis added.

53. John Durham Peters, "Calendar, Clock, Tower," in *Deus in Machina: Religion and Technology in Historical Perspective,* ed. Jeremy Stolow (New York: Fordham University Press, 2013), 25–42; John Durham Peters, *The Marvelous Clouds: Toward a Philosophy of Elemental Media* (Chicago: University of Chicago Press, 2015); Joshua Reeves and Jeremy Packer, "Police Media: The Governance of Territory, Speed, and Communication," *Communication and Critical/Cultural Studies* 10, no. 4 (2013): 359–84; Ned Rossiter, "Locative Media as Logistical Media: Situating Infrastructure and the Governance of Labor in Supply-Chain Capitalism," in *Locative Media,* ed. Rowan Wilken and Gerard Goggin (New York: Routledge, 2014), 208–21.

54. Peters, "Calendar, Clock, Tower," 41.

55. Rossiter, *Software, Infrastructure, Labor,* 5.

56. See also Shannon Mattern, *Code and Clay, Data and Dirt: Five Thousand Years of Urban Media* (Minneapolis: University of Minnesota Press, 2017).

57. Peters, *Marvelous Clouds,* 23.

58. Liam Cole Young, "Cultural Techniques and Logistical Media: Tuning German and Anglo-American Media Studies," *M/C Journal* 18, no. 2 (2015), http://journal.media-culture.org.au/index.php/mcjournal/article/view/961.

59. Mark Andrejevic, Alison Hearn, and Helen Kennedy, "Cultural Studies of Data Mining: Introduction," *European Journal of Cultural Studies* 18, no. 4 (2015): 379–94.

60. Julian Reid, *The Biopolitics of the War on Terror: Life Struggles, Liberal Modernity, and the Defense of Logistical Societies* (Manchester, UK: Manchester University Press, 2006), 13.

61. Brian Larkin, "The Politics and Poetics of Infrastructure," *Annual Review of Anthropology* 42, no. 1 (2013): 327–43.

62. Andrés Luque-Ayala and Simon Marvin, "The Maintenance of Urban Circulation: An Operational Logic of Infrastructural Control," *Environment & Planning D: Society & Space* 34, no. 2 (2016): 191–208.

63. James W. Carey, "Technology and Ideology: The Case of the Telegraph," in *Communication as Culture: Essays on Media and Society* (New York: Psychology Press, 1989), 201–29.

64. AbdouMaliq Simone, "The Surfacing of Urban Life," *City* 15, nos. 3–4 (2011): 356, emphasis added. See also AbdouMaliq Simone, "People as Infrastructure: Intersecting Fragments in Johannesburg," *Public Culture* 16, no. 3 (2004): 407–29.

65. Rossiter, *Software, Infrastructure, Labor*, 5.

66. Herbert Marcuse, "Some Social Implications of Modern Technology," in *The Essential Frankfurt School Reader*, ed. Andrew Arato and Eike Gebhardt (New York: Continuum, 1977), 138–62.

67. Jathan Sadowski and Frank Pasquale, "The Spectrum of Control: A Social Theory of the Smart City," *First Monday* 20, no. 7 (2015), https://doi.org/10.5210/fm.v20i7.5903; Jennifer Gabrys, "Programming Environments: Environmentality and Citizen Sensing in the Smart City," *Environment & Planning D: Society & Space* 32 (2014): 30–48; Tim Harford, "Big Data: A Big Mistake?" *Significance* 11, no. 5 (2014): 14–19; Jim Thatcher, "Avoiding the Ghetto through Hope and Fear: An Analysis of Immanent Technology Using Ideal Types," *GeoJournal* 78, no. 6 (2013): 967–80; Jim Thatcher, "Living on Fumes: Digital Footprints, Data Fumes, and the Limitations of Spatial Big Data," *International Journal of Communication* 8 (2014): 1765–83; Louise Amoore, "Biometric Borders: Governing Mobilities in the War on Terror," *Political Geography* 25, no. 3 (2006): 336–51; Louise Amoore, "Data Derivatives: On the Emergence of. Security Risk Calculus for Our Times," *Theory, Culture & Society* 28, no. 6 (2011): 24–43; Kelly A. Gates, *Our Biometric Future: Facial Recognition Technology and the Culture of Surveillance* (New York: NYU Press, 2011).

68. Kanngieser, "Tracking and Tracing," 595.

69. Marcuse, "Some Social Implications of Modern Technology"; Herbert Marcuse, *One-Dimensional Man: Studies in the Ideology of Advanced Industrial Society* (Boston: Beacon Press, 2012).

70. Jacques Ellul, *The Technological Society*, trans. John Wilkinson (New York: Vintage Books, 1964), 5.

71. Ellul, 5.

72. Ellul, 6.

73. Michael Batty, "How Disruptive Is the Smart Cities Movement?" *Environment & Planning B: Urban Analytics & City Science* 43, no. 3 (May 2016): 441–43.

74. Francesco Paolo Appio, Marcos Lima, and Sotirios Paroutis, "Understanding Smart Cities: Innovation Ecosystems, Technological Advancements, and Societal Challenges," *Technological Forecasting and Social Change* 142 (May 2019): 1–14; Ellie Cosgrave, Kate Arbuthnot, and Theo Tryfonas, "Living Labs, Innovation Districts and Information Marketplaces: A Systems Approach for Smart Cities," *Procedia Computer Science* 16 (January 2013): 668–77; Carlo Ratti and Matthew Cladel, "Life in the Uber City," *Design Indaba*, July 14, 2014, http://www.designindaba.com/articles/point-view/opinion-life-uber-city.

75. Federico Cugurullo, "The Origin of the Smart City Imaginary: From the Dawn of Modernity to the Eclipse of Reason," in *The Routledge Companion to Urban Imaginaries*, ed. Christopher Lindner and Miriam Meissner (New York: Routledge, 2019), 113–24; Halpern, *Beautiful Data*; Mattern, *Code and Clay*; Shelton, Zook, and Wiig, "The 'Actually Existing Smart City'"; Zook, "Crowd-Sourcing the Smart City."

76. Michel Foucault, *Security, Territory, Population: Lectures at the College de France, 1977–78*, trans. Graham Burchell (New York: Palgrave Macmillan, 2007).

77. Anthony Townsend, "Cities of Data: Examining the New Urban Science," *Public Culture* 27, no. 2 (2015): 203.

78. Piers Beirne, *Inventing Criminology: Essays on the Rise of "Homo Criminalis"* (Albany: State University of New York Press, 1993); Theo Kindynis, "Ripping Up the Map Criminology and Cartography Reconsidered," *British Journal of Criminology* 54, no. 2 (2014): 222–43.

79. Bernard E. Harcourt, *Against Prediction: Profiling, Policing, and Punishing in an Actuarial Age* (Chicago: University of Chicago Press, 2006).

80. Craig M. Dalton and Jim Thatcher, "Inflated Granularity: Spatial 'Big Data' and Geodemographics," *Big Data & Society* 2, no. 2 (2015): 1–15; Liz Mcfall, *Devising Consumption: Cultural Economics of Insurance, Credit and Spending* (New York: Routledge, 2015); Dan Bouk, *How Our Days Became Numbered: Risk and the Rise of the Statistical Individual* (Chicago: University of Chicago Press, 2015); Josh Lauer, *Creditworthy: A History of Consumer Surveillance and Financial Identity in America* (New York: Columbia University Press, 2017).

81. Manuel B. Aalbers, "Do Maps Make Geography? Part 1: Redlining, Planned Shrinkage, and the Places of Decline," *ACME: An International Journal for Critical Geographies* 13, no. 4 (2014): 525–56; Beryl Satter, *Family Properties: How the Struggle over Race and Real Estate Transformed Chicago and Urban America* (New York: Picador, 2010); Keeanga-Yamahtta Taylor, *Race for Profit: How Banks and*

the Real Estate Industry Undermined Black Homeownership (Chapel Hill: University of North Carolina Press, 2019); Jennifer S. Light, *From Warfare to Welfare: Defense Intellectuals and Urban Problems in Cold War America* (Baltimore: Johns Hopkins University Press, 2003).

82. R. E. Klosterman and J. D. Landis, "Microcomputers in US Planning: Past, Present, and Future," *Environment & Planning B: Planning & Design* 15, no. 3 (1988): 355–67; Emile Tanic, "Urban Planning and Artificial Intelligence: The Urbys System," *Computers, Environment & Urban Systems* 10, no. 3 (1986): 135–46.

83. Mark Weiser, "The Computer for the 21st Century," *Scientific American* 265, no. 3 (1991): 94–104; Paul Dourish and Genevieve Bell, *Divining a Digital Future: Mess and Mythology in Ubiquitous Computing* (Cambridge, Mass.: MIT Press, 2011); Stephen Graham, "Introduction: From Dreams of Transcendence to the Remediation of Urban Life," in *The Cybercities Reader*, ed. Stephen Graham (New York: Routledge, 2003), 1–29.

84. Cugurullo, "Origin of the Smart City Imaginary"; Mark Shepard, *Sentient City: Ubiquitous Computing, Architecture, and the Future of Urban Space* (New York: Architectural League of New York, 2011).

85. Lewis Mumford, *The City in History: Its Origins, Its Transformations, and Its Prospects* (New York: Mariner Books, 1968); Lewis Mumford, *Technics and Civilization* (1931; repr., Chicago: University of Chicago Press, 2010).

86. Hollands, "Will the Real Smart City Please Stand Up?"; Vito Albino, Umberto Berardi, and Rosa Maria Dangelico, "Smart Cities: Definitions, Dimensions, Performance, and Initiatives," *Journal of Urban Technology* 22, no. 1 (2015): 3–21.

87. Donald McNeill, "Global Firms and Smart Technologies: IBM and the Reduction of Cities," *Transactions of the Institute of British Geographers* 40, no. 4 (2015): 562–74.

88. Stephen Graham and Simon Marvin, *Splintering Urbanism: Networked Infrastructures, Technological Mobilities and the Urban Condition* (New York: Routledge, 2001).

89. Mark Blyth, *Austerity: The History of a Dangerous Idea* (New York: Oxford University Press, 2015); Nick Srnicek, *Platform Capitalism* (Malden, Mass.: Polity Press, 2016).

90. Donatella della Porta et al., *Movement Parties against Austerity* (Malden, Mass.: Polity Press, 2017); David Harvey, "Neoliberalism and the City," *Studies in Social Justice* 1, no. 1 (2007): 2–13; David Harvey, *A Brief History of Neoliberalism* (Oxford, UK: Oxford University Press, 2007).

91. James Ferguson, *Global Shadows: Africa in the Neoliberal World Order* (Durham, N.C.: Duke University Press, 2006); David Harvey, "Neoliberalism as Creative Destruction," *Annals of the American Academy of Political and Social Science* 610, no. 1 (2007): 21–44; Harvey, *Brief History of Neoliberalism*; Mathieu Hilgers, "The Historicity of the Neoliberal State," *Social Anthropology* 20, no. 1 (2012): 80–94.

92. Jan Nederveen Pieterse, "Structural Adjustment," *The Futures We Want: Global Sociology and the Struggles for a Better World,* August 19, 2015, http://futureswewant.net/jan-nederveen-pieterse-europe/.

93. Srnicek, *Platform Capitalism,* 31.

94. Fiona Allon and Guy Redden, "The Global Financial Crisis and the Culture of Continual Growth," *Journal of Cultural Economy* 5, no. 4 (November 2012): 375–90; Jason Glynos, Robin Klimecki, and Hugh Willmott, "Cooling Out the Marks: The Ideology and Politics of the Financial Crisis," *Journal of Cultural Economy* 5, no. 3 (August 2012): 297–320; Adam Tooze, *Crashed: How a Decade of Financial Crises Changed the World* (New York: Viking Press, 2018).

95. Jamie Peck, "Austerity Urbanism," *City* 16, no. 6 (2012): 626–55; Jamie Peck, Theodore Nik, and Brenner Neil, "Neoliberal Urbanism Redux?" *International Journal of Urban and Regional Research* 37, no. 3 (2013): 1091–99.

96. Peck, "Austerity Urbanism," 630.

97. For example, see Edward Glaeser, *Triumph of the City: How Our Greatest Invention Makes Us Richer, Smarter, Greener, Healthier, and Happier* (New York: Penguin Books, 2011).

98. Quoted in David Harvey, "Flexible Accumulation through Urbanization: Reflections on 'Post-Modernism' in the American City," *Perspecta* 26 (1990): 251–72.

99. Thomas Osborne and Nikolas Rose, "Governing Cities: Notes on the Spatialization of Virtue," *Environment & Planning D: Society & Space* 17 (1999): 737–60; Mary Poovey, *Making a Social Body: British Cultural Formation, 1830–1864* (Chicago: University of Chicago Press, 1995).

100. David Harvey, "From Managerialism to Entrepreneurialism: The Transformation in Urban Governance in Late Capitalism," *Geografiska Annaler. Series B. Human Geography* 71 (1989): 3–17.

101. Steven Conn, *Americans against the City: Anti-Urbanism in the Twentieth Century* (Oxford, UK: Oxford University Press, 2014).

102. Thomas J. Sugrue, *The Origins of the Urban Crisis: Race and Inequality in Postwar Detroit* (Princeton, N.J.: Princeton University Press, 2014); Jodie Adams Kirshner and Michael Eric Dyson, *Broke: Hardship and Resilience in a City of Broken Promises* (New York: St. Martin's Press, 2019).

103. Naomi Klein, *The Shock Doctrine: The Rise of Disaster Capitalism* (New York: Picador, 2007).

104. Quoted in Peck, "Austerity Urbanism," 628.

105. Andrea Pollio, "Technologies of Austerity Urbanism: The 'Smart City' Agenda in Italy (2011–2013)," *Urban Geography* 37, no. 4 (May 2016): 514–34.

106. Wiig, "Empty Rhetoric of the Smart City."

107. Wim Dierckxsens, *The Limits of Capitalism: An Approach to Globalization without Neoliberalism* (London: Zed Books, 2001); David Harvey, *Spaces of Global*

Capitalism: A Theory of Uneven Geographical Development (New York: Verso, 2006).

108. Ash Amin and Kevin Robins, "The Re-Emergence of Regional Economies? The Mythical Geography of Flexible Accumulation," *Environment & Planning D: Society & Space* 8, no. 1 (1990): 7–34; Harvey, "Flexible Accumulation through Urbanization"; Erica Schoenberger, "From Fordism to Flexible Accumulation: Technology, Competitive Strategies, and International Location," *Environment & Planning D: Society & Space* 6, no. 3 (1988): 245–62.

109. David Harvey, *The Limits to Capital* (London: Verso, 2018); Judith Stein, *Pivotal Decade: How the United States Traded Factories for Finance in the Seventies* (New Haven, Conn.: Yale University Press, 2010).

110. Schoenberger, "From Fordism to Flexible Accumulation."

111. Schoenberger, 252.

112. Shoshana Zuboff, *In the Age of the Smart Machine: The Future of Work and Power* (New York: Basic Books, 1988), 9.

113. Jathan Sadowski, "When Data Is Capital: Datafication, Accumulation, and Extraction," *Big Data & Society* 6, no. 1 (2019), https://doi.org/10.1177/2053951718820549; Richard Salame, "The New Taylorism," *Jacobin*, February 20, 2018, https://jacobinmag.com/2018/02/amazon-wristband-surveillance-scientific-management; Graham Sewell and Barry Wilkinson, "'Someone to Watch Over Me': Surveillance, Discipline and the Just-in-Time Labour Process," *Sociology* 26, no. 2 (1992): 271–89.

114. George Stalk, Jr., "Time: The Next Source of Competitive Advantage," *Harvard Business Review*, July 1, 1988, https://hbr.org/1988/07/time-the-next-source-of-competitive-advantage.

115. Beniger, *Control Revolution*; Chandler, *Visible Hand*.

116. J. Ben Naylor, Mohamed M. Naim, and Danny Berry, "Leagility: Integrating the Lean and Agile Manufacturing Paradigms in the Total Supply Chain," *International Journal of Production Economics* 62, no. 1 (1999): 107–18; Denis R. Towill and Martin Christopher, "Supply Chain Migration from Lean and Functional to Agile and Customised," *Supply Chain Management* 5, no. 4 (2000): 206–13.

117. Richard Langlois, "The Vanishing Hand: The Changing Dynamics of Industrial Capitalism," *Economics Working Papers* 200221 (2002), http://digitalcommons.uconn.edu/econ_wpapers/200221; Richard Langlois, "Vertical (Dis)integration and Technological Change," *Organizations and Markets*, May 18, 2012, https://organizationsandmarkets.com/2012/05/18/vertical-disintegration-and-technological-change/.

118. Chua et al., "Introduction," 619; Tsing, "Supply Chains and the Human Condition"; Andrew Sayer, "New Developments in Manufacturing: The Just-in-Time System," *Capital & Class* 10, no. 3 (1986): 43–72.

119. Lyster, *Learning from Logistics*.

120. Danyluk, "Capital's Logistical Fix."

121. Peter Drucker, "The Economy's Dark Continent," *Fortune,* April 1962, 103.

122. Kanngieser, "Tracking and Tracing"; Rossiter, *Software, Infrastructure, Labor*; Rossiter, "Locative Media"; Adam Robinson, "The Evolution and History of Supply Chain Management," *Cerasis: Transportation Management,* January 23, 2015, https://cerasis.com/history-of-supply-chain-management/.

123. Kenneth T. Jackson, *Crabgrass Frontier: The Suburbanization of the United States* (Oxford, UK: Oxford University Press, 1987); Lynn Spigel, *Welcome to the Dreamhouse: Popular Media and Postwar Suburbs* (Durham, N.C.: Duke University Press, 2001).

124. Kevin M. Kruse, *White Flight: Atlanta and the Making of Modern Conservatism* (Princeton, N.J.: Princeton University Press, 2013); Sugrue, *Origins of the Urban Crisis.*

125. Harvey, "Flexible Accumulation through Urbanization."

126. Mimi Sheller and John Urry, eds., *Mobile Technologies of the City* (New York: Routledge, 2012).

127. U.S. Department of Transportation, "Smart City Challenge: Urban Freight Delivery and Logistics," January 6, 2016, https://www.transportation.gov/smart city/infosessions/freight-logistics.

128. Mark Graham, Rob Kitchin, Shannon Mattern, and Joe Shaw, eds., *How to Run a City Like Amazon, and Other Fables* (Oxford, UK: Meat Space Press, 2019); Stephen Goldsmith and Neil Kleiman, "Cities Should Act More Like Amazon to Better Serve Their Citizens," *NextCity*, January 23, 2018, https://nextcity.org/daily/entry/cities-should-act-more-like-amazon-to-better-serve-their-citizens; see also Ingrid Burrington, "What Amazon Taught the Cops," *The Nation,* May 27, 2015, https://www.thenation.com/article/what-amazon-taught-cops/; Emily West, "Amazon: Surveillance as a Service," *Surveillance & Society* 17, nos. 1–2 (2019): 27–33.

129. Torin Monahan, "The Image of the Smart City: Surveillance Protocols and Social Inequality," in *Handbook of Cultural Security,* ed. Yasushi Watanabe (Cheltenham, UK: Edward Elgar), 210–26; Reid, *Biopolitics of the War on Terror*; Stephen Graham, *Cities under Siege: The New Military Urbanism* (New York: Verso, 2014); David Lyon, ed., *Surveillance as Social Sorting: Privacy, Risk and Automated Discrimination* (New York: Routledge, 2003).

130. Zygmunt Bauman and David Lyon, *Liquid Surveillance* (Malden, Mass.: Polity Press, 2013); Chua et al., "Introduction"; Cowen, *Deadly Life of Logistics*; Lisa Parks, "Points of Departure: The Culture of US Airport Screening," *Journal of Visual Culture* 6, no. 2 (2007): 183–200; Graham, *Cities under Siege*; Neilson, "Five Theses"; Lisa Parks, "Plotting the Personal: Global Positioning Satellites and Interactive Media," *Ecumene* 8, no. 2 (2001): 209–22; Ben Anderson, "Becoming and Being Hopeful: Towards a Theory of Affect," *Environment & Planning D: Society & Space* 24, no. 5 (2006): 733–52; Mark Andrejevic, "The Work That Affective Economics Does," *Cultural Studies* 25, nos. 4–5 (2011): 604–20; Adam

Arvidsson and Elanor Colleoni, "Value in Informational Capitalism and on the Internet," *The Information Society* 28, no. 3 (2012): 135–50; Nigel Thrift, "Lifeworld Inc—and What to Do about It," *Environment & Planning D: Society & Space* 29, no. 1 (2011): 5–26.

131. Colin J. Bennett and Priscilla M. Regan, "Surveillance and Mobilities," *Surveillance & Society* 1, no. 4 (2003), https://doi.org/10.24908/ss.v1i4.3330; Jennie Germann Molz, "'Watch Us Wander': Mobile Surveillance and the Surveillance of Mobility," *Environment & Planning A* 38, no. 2 (2006): 377–93; David Lyon, "Surveillance Studies: Understanding Visibility, Mobility, and the Phenetic Fix," *Surveillance & Society* 1, no. 1 (2002): 1–7.

132. Ryan Bishop and John W. P. Phillips, "The Urban Problematic," *Theory, Culture & Society* 30, nos. 7–8 (2013): 221–41; Ryan Bishop and John W. P. Phillips, "The Urban Problematic II," *Theory, Culture & Society* 31, nos. 7–8 (2014): 121–36.

133. Lyon, "Surveillance Studies," 3.

134. Loïc Wacquant, "Three Steps to a Historical Anthropology of Actually Existing Neoliberalism," *Social Anthropology* 20, no. 1 (2012): 74. See also Steve Herbert and Elizabeth Brown. "Conceptions of Space and Crime in the Punitive Neoliberal City." *Antipode* 38, no. 4 (2006): 755–77.

135. Pete Fussey, "Command, Control and Contestation: Negotiating Security at the London 2012 Olympics," *The Geographical Journal* 181, no. 3 (2015): 214.

136. Peter Marcuse, "The Enclave, the Citadel, and the Ghetto: What Has Changed in the Post-Fordist US City," *Urban Affairs Review* 33, no. 2 (1997): 228–64.

137. Marcuse, 228.

138. Philippe Bourgois and Jeffrey Schonberg, *Righteous Dopefiend* (Berkeley: University of California Press, 2009); Loïc Wacquant, *Urban Outcasts: A Comparative Sociology of Advanced Marginality* (Malden, Mass.: Polity Press, 2007); Loïc Wacquant, *Punishing the Poor: The Neoliberal Government of Social Insecurity* (Durham, N.C.: Duke University Press, 2009).

139. Alan Wiig, "Secure the City, Revitalize the Zone: Smart Urbanization in Camden, New Jersey," *Environment & Planning C: Politics & Space* 36, no. 3 (2018): 403–22.

140. Lyon, "Surveillance Studies"; Lorna Muir, "Control Space? Cinematic Representations of Surveillance Space between Discipline and Control," *Surveillance & Society* 9, no. 3 (2012): 263–79.

141. Thomas Nail, *Theory of the Border* (Oxford, UK: Oxford University Press, 2015).

142. Sterling, "Stop Saying 'Smart Cities.'"

143. Sarah Sharma, "Taxis as Media: A Temporal Materialist Reading of the Taxi-Cab," *Social Identities* 14, no. 4 (2008): 457–64; Sarah Sharma, *In the Meantime: Temporality and Cultural Politics* (Durham, N.C.: Duke University Press, 2014).

144. Brian Jordan Jefferson, "Digitize and Punish: Computerized Crime Mapping and Racialized Carceral Power in Chicago," *Environment & Planning D: Society & Space* 35, no. 5 (2017): 775–96.

145. Thrift, "Lifeworld Inc.," 11.

146. Sarah Sharma, "Baring Life and Lifestyle in the Non-Place," *Cultural Studies* 23, no. 1 (2009): 129–48.

147. Mezzadra and Neilson, "On the Multiple Frontiers of Extraction."

148. Steve Hamilton and Ximon Zhu, *Funding and Financing Smart Cities* (Washington, D.C.: Deloitte Center for Government Insights, 2017), https://www2.deloitte.com/content/dam/Deloitte/us/Documents/public-sector/us-ps-funding-and-financing-smart-cities.pdf.

149. "CityBridge to Launch LinkNYC, Largest Urban Digital Ad Network," *Screen Media Daily,* 2014, http://screenmediadaily.com/citybridge-to-launch-linknyc-largest-urban-digital-ad-network/ (site discontinued).

150. Lee McGuigan, "Automating the Audience Commodity: The Unacknowledged Ancestry of Programmatic Advertising," *New Media & Society* 21, nos. 11–12 (2019): 2366–85.

151. Bernes, "Logistics, Counterlogistics and the Communist Prospect."

152. Alberto Toscano, "Logistics and Opposition," *Mute,* August 9, 2011, http://www.metamute.org/editorial/articles/logistics-and-opposition.

153. Elijah Anderson, *Code of the Street: Decency, Violence, and the Moral Life of the Inner City* (New York: W. W. Norton & Company, 2000), 15.

154. Inga Saffron, "Searching for a New Identity, Philadelphia's Most Gentrified Neighborhood Looks to Its African American Past," *Philadelphia Inquirer,* July 19, 2018, https://www.inquirer.com/philly/columnists/inga_saffron/philadelphia-neighborhood-names-graduate-hospital-marian-anderson-20180719.html.

155. Caitlin McCabe, "Councilman Kenyatta Johnson Pushes Ban on Bay Windows, Seen as a Symbol of Gentrification, in His South Philly District," *Philadelphia Inquirer,* May 28, 2019, https://www.inquirer.com/real-estate/housing/councilman-kenyatta-johnson-bay-windows-balconies-bill-point-breeze-south-philly-development-gentrification-20190528.html.

156. "Philly Tech Week," Technical.ly, https://technical.ly/project/philly-tech-week/.

157. Daryl C. Murphy, "Poverty Still Plaguing Philadelphia, Poorest Big City in the Country," *WHYY,* September 14, 2018, https://whyy.org/articles/poverty-still-plaguing-philadelphia-poorest-big-city-in-the-country/.

158. Jennifer Percy, "Trapped by the 'Walmart of Heroin,'" *New York Times,* October 10, 2018, https://www.nytimes.com/2018/10/10/magazine/kensington-heroin-opioid-philadelphia.html.

159. Alfred Lubrano, "New Census Figures on Philly Neighborhoods Show Inequality, High Numbers of Whites Living in Poverty," *Philadelphia Inquirer,*

December 6, 2018, https://www.inquirer.com/philly/news/poverty-new-census-figures-philadelphia-neighborhoods-whites-opioids-20181206.html-2.

160. "By the Numbers: Pennovation Works' Ascendant 2018," *Penn Today,* January 9, 2019, https://penntoday.upenn.edu/news/numbers-pennovation-works-ascendant-2018.

161. "About the Works," Pennovation Works, October 1, 2016, https://www.pennovation.upenn.edu/about-the-works.

162. Jacob Adelman, "Refinery Shutdown Could Mean New Life for 1,400 Waterfront Acres," *Philadelphia Inquirer,* June 26, 2019, https://www.inquirer.com/news/philadelphia-refinery-complex-real-estate-land-use-port-housing-remediation-20190626.html.

163. Delaware Valley Regional Planning Commission, "Future Forces 2050 Working Group Meeting 1" (PowerPoint presentation, Greater Philadelphia Futures Working Group, February 15, 2019), https://www.dvrpc.org/LongRangePlan/FuturesGroup/pdf/2.15.19_Presentation.pdf.

1. Design

1. Robin Goldwyn Blumenthal, "Endgame for the Pay Phone," *Barron's,* 2014, http://www.barrons.com/articles/endgame-for-the-pay-phone-1416630110.

2. World Bank, "Mobile Cellular Subscriptions (per 100 People)," 2016, http://data.worldbank.org/indicator/IT.CEL.SETS.P2?locations=US.

3. Pew Research Center, "Demographics of Mobile Device Ownership and Adoption in the United States," June 12, 2019, https://www.pewinternet.org/fact-sheet/mobile/.

4. Matthew Flamm, "Calling Back the Pay Phone," *Crain's New York Business,* 2012, http://www.crainsnewyork.com/article/20121118/TECHNOLOGY/311189976/calling-back-the-pay-phone.

5. For an example of artist "hacks" of payphones, see Rozemarijn Stam, "NYC Phone Booths Become Art Galleries," *Pop-Up City,* April 24, 2018, https://popupcity.net/nyc-phone-booths-become-art-galleries/. Mark A. Thomas's Payphone Project (https://www.payphone-project.com/) has been gathering and collating documentation of payphones from around the world since 1995.

6. "Sandy Aftermath: Reliable Pay Phones You Can Use," *Bowery Boogie,* November 1, 2012, https://www.boweryboogie.com/2012/11/sandy-aftermath-reliable-pay-phones-you-can-use/.

7. Devindra Hardawar, "Mayor Bloomberg Calls upon Techies to Reinvent New York City's Payphones," *VentureBeat,* December 4, 2012, https://venturebeat.com/2012/12/04/bloomberg-reinvent-payphones-nyc/.

8. Kevin Rogan, "What's a Smart City Supposed to Look Like?" *CityLab,* June 27, 2019, https://www.citylab.com/design/2019/06/smart-city-photos-technology-marketing-branding-jibberjabber/592123/.

9. "CityBridge to Launch LinkNYC, Largest Urban Digital Ad Network," *Screen Media Daily,* 2014, http://screenmediadaily.com/citybridge-to-launch-linknyc-largest-urban-digital-ad-network/ (site discontinued).

10. Aaron Shapiro, "The True Cost of Free LinkPHL Wi-Fi Might Be Privacy," *Philadelphia Inquirer,* January 9, 2019, https://www.inquirer.com/opinion/commentary/link-phl-wifi-kiosks-data-privacy-20190109.html; Ingrid Lunden, "LinkNYC's Free WiFi and Phone Kiosks Hit London as LinkUK, in Partnership with BT," *TechCrunch,* October 25, 2016, http://social.techcrunch.com/2016/10/25/linknycs-free-wifi-and-phone-kiosks-hit-london-as-linkuk-in-partnership-with-bt/.

11. The VHS–Betamax race is a canonical example of technologies competing to define the standard; see Joshua M. Greenberg, *From Betamax to Blockbuster: Video Stores and the Invention of Movies on Video* (Cambridge, Mass.: MIT Press, 2010). On the theoretical foundations of the sociology of expectations, see Carla Alvial-Palavicino, "The Future as Practice: A Framework to Understand Anticipation in Science and Technology," *Tecnoscienza: The Italian Journal of Science & Technology Studies* 6, no. 2 (2016): 135–72; Mads Borup et al., "The Sociology of Expectations in Science and Technology," *Technology Analysis & Strategic Management* 18, nos. 3–4 (2006): 285–98; Nik Brown and Mike Michael, "A Sociology of Expectations: Retrospecting Prospects and Prospecting Retrospects," *Technology Analysis & Strategic Management* 15, no. 1 (2003): 3–18; Nik Brown, Brian Rappert, and Andrew Webster, *Contested Futures: A Sociology of Prospective Techno-Science* (Farnham, UK: Ashgate, 2000); Anne Galloway, "Locating Media Futures in the Present: Or How to Map Emergent Associations and Expectations," *Aether: The Journal of Media Geography* 5 (2010): 27–36; Anne Galloway, "Emergent Media Technologies, Speculation, Expectation, and Human/Nonhuman Relations," *Journal of Broadcasting & Electronic Media* 57, no. 1 (2013): 53–65.

12. Michael Provenzano, "Why Google Needs to Focus on Connected Cities for Its Alphabet Strategy to Pay Off," *Smart Cities Dive,* July 12, 2017, http://www.smartcitiesdive.com/news/why-google-needs-to-focus-on-connected-cities-for-its-alphabet-strategy-to/446908/.

13. Outdoor Advertising Association of America, "OOH Growth Continues amid Traditional Media Declines," May 20, 2019, https://oaaa.org/portals/0/Public%20PDFs/Outlook%20Newsletter/OutdoorOutlook_052019.pdf.

14. Outdoor Adverting Association of America, "Industry Revenue," March 22, 2018. https://oaaa.org/StayConnected/NewsArticles/IndustryRevenue.aspx; Alison Weissbrot, "4 Things to Know as Out-Of-Home Goes Programmatic," *AdExchanger,* February 15, 2017, https://adexchanger.com/digital-out-of-home/4-things-know-home-goes-programmatic/.

15. "Peek-a-Boo, Google Sees You," *CNN Money,* April 7, 2006, https://money.cnn.com/2006/04/06/technology/googsf_reut/.

16. Sydney Ember, "See That Billboard? It May See You, Too," *New York Times,* February 29, 2016, https://www.nytimes.com/2016/02/29/business/media/see-that-billboard-it-may-see-you-too.html.

17. Orit Halpern, Jesse LeCavalier, Nevea Calvillo, and Wolfgang Pietsch, "Test-Bed Urbanism." *Public Culture* 25, no. 2 (2013): 272–306; Mark Andrejevic, and Mark Burdon. "Defining the Sensor Society." *Television & New Media* 16, no. 1 (2015): 20.

18. Craig Campbell, "LinkNYC Kiosks Provide Free Internet, but for a Price," *Government Technology,* February 9, 2016, http://www.govtech.com/dc/articles/LinkNYC-Kiosks-Provide-Free-Internet-But-for-a-Price.html; Sam Gustin, "LinkNYC's New Free Network Is Blazing Fast: But at What Cost to Privacy?," *Motherboard,* February 19, 2016, http://motherboard.vice.com/read/linknycs-new-free-network-is-blazing-fast-but-at-what-cost-to-privacy; Ava Kofman, "Are New York's Free LinkNYC Internet Kiosks Tracking Your Movements?" *The Intercept,* September 8, 2018, https://theintercept.com/2018/09/08/linknyc-free-wifi-kiosks/.

19. Arman Tabatabai, "The Economics and Trade-Offs of Ad-Funded Smart City Tech," *TechCrunch,* December 1, 2018, http://social.techcrunch.com/2018/12/01/the-economics-and-tradeoffs-of-ad-funded-smart-city-tech/.

20. Quoted in eMarketer, "Listen In: Blending Tech with Media to Make Cities Smarter," *Behind the Numbers* (podcast), 2016, https://www.emarketer.com/Article/Listen-In-Blending-Tech-with-Media-Make-Cities-Smarter/1014312.

21. Mark Andrejevic, "Surveillance in the Digital Enclosure," *Communication Review* 10, no. 4 (2007): 295–317; Mark Andrejevic, "Privacy, Exploitation, and the Digital Enclosure," *Amsterdam Law Forum* 1, no. 4 (2009): 47–62; Lee McGuigan, "Automating the Audience Commodity: The Unacknowledged Ancestry of Programmatic Advertising," *New Media & Society* 21, nos. 11–12 (November 2019): 2366–85; Jeremy Packer, "Epistemology Not Ideology, or Why We Need New Germans," *Communication & Critical/Cultural Studies* 10, nos. 2–3 (2013): 295–300; Joseph Turow and Nick Couldry, "Media as Data Extraction: Towards a New Map of a Transformed Communications Field," *Journal of Communication* 68, no. 2 (2018): 415–23; Jose van Dijck, "Datafication, Dataism and Dataveillance: Big Data between Scientific Paradigm and Ideology," *Surveillance & Society* 12, no. 2 (2014): 197–208.

22. eMarketer, "Listen In."

23. Quoted in Nick Pinto, "Google Is Transforming NYC's Payphones into a 'Personalized Propaganda Engine,'" *Village Voice,* July 6, 2016, https://www.villagevoice.com/2016/07/06/google-is-transforming-nycs-payphones-into-a-personalized-propaganda-engine/.

24. Pinto, "Google Is Transforming NYC's Payphones."

25. Quoted in Pinto.

26. Brown and Michael, "A Sociology of Expectations"; Tomas Moe Skjølsvold, "Back to the Futures: Retrospecting the Prospects of Smart Grid Technology," *Futures* 63 (2014): 26–36; Nora A Draper and Joseph Turow, "The Corporate Cultivation of Digital Resignation," *New Media & Society* 21, no. 8 (2019): 1824–39; Lisa Parks, "'Stuff You Can Kick': Toward a Theory of Media Infrastructures," in *Between Humanities and the Digital,* ed. Patrik Svensson and David Theo Goldberg (Cambridge, Mass.: MIT Press, 2015), 355–74.

27. Alexander R. Galloway, *The Interface Effect* (Malden, Mass.: Polity, 2012).

28. Shannon Mattern, "Interfacing Urban Intelligence," *Places Journal,* April 2014, https://placesjournal.org/article/interfacing-urban-intelligence/.

29. Mark Memmott, "Pay Phones are Suddenly Important Again Because of Sandy," *NPR,* November 1, 2012, https://www.npr.org/sections/thetwo-way/2012/11/01/164108616/pay-phones-are-suddenly-important-again-because-of-sandy; Ben Cohen, "After Sandy, Wired New Yorkers Get Reconnected with Pay Phones," *Wall Street Journal,* November 1, 2012, https://www.wsj.com/articles/SB10001424052970203707604578091141769109974.

30. Will Oremus, "Michael Bloomberg Wants to Do Something Cool with New York's Phone Booths," *Slate,* December 5, 2012, http://www.slate.com/blogs/future_tense/2012/12/05/reinvent_payphones_new_york_challenges_techies_to_design_a_better_phone.html.

31. Quoted in Sam Byford, "New York City Launches Competition to Create the Perfect Payphone," *The Verge,* December 4, 2012, https://www.theverge.com/2012/12/4/3729950/new-york-city-payphone-design-competition.

32. Hardawar, "Mayor Bloomberg Calls upon Techies."

33. Nick Sbordone (@nicksbordone), "I have never, and I mean NEVER been as amped about payphones as I am at this very moment. It's on. Come and #ReinventPayphones," Twitter, December 4, 2012, 7:19 p.m., https://twitter.com/nicksbordone/status/276118598785835008.

34. NYC Office of the Mayor, "Mayor Bloomberg, Chief Information and Innovation Officer Merchant and Chief Digital Officer Haot Launch Reinvent Payphones Design Challenge," December 4, 2012, http://www.nyc.gov/ (specific page discontinued).

35. NYC.gov, "Re-Invent Payphones Information Session," January 23, 2013 (YouTube video, uploaded February 1, 2013), https://www.youtube.com/watch?v=J5IoW8mYSjk.

36. NYC.gov video.

37. Rachel Sterne Haot (@rachelhaot), "In the #ReinventPayphones Design Challenge - from wifi to digital kiosks - the sky is the limit. Go beyond the phone: http://nyc.gov/reinventpayphones," Twitter, December 4, 2012, 6:48 p.m., https://twitter.com/nicksbordone/status/276118598785835008.

38. NYC.gov video.

39. Mariana Valverde, *Everyday Law on the Street: City Governance in an Age of Diversity* (Chicago: University of Chicago Press, 2012).

40. NYC.gov, "Re-Invent Payphones Information Session."

41. NYC.gov video.

42. David Harvey, "From Managerialism to Entrepreneurialism: The Transformation in Urban Governance in Late Capitalism," *Geografiska Annaler, Series B: Human Geography* 71, no. 1 (1989): 3–17.

43. Harvey; see also Harvey Molotch, "The City as a Growth Machine: Toward a Political Economy of Place," *American Journal of Sociology* 82, no. 2 (1976): 309–32.

44. Sharon Zukin, "The Origins and Perils of Development in the Urban Tech Landscape," *The Architect's Newspaper,* May 8, 2019, https://archpaper.com/2019/05/urban-tech-landscape/.

45. Sarah Holder, "What Did Cities Actually Offer Amazon?" *CityLab,* May 29, 2018, https://www.citylab.com/life/2018/05/what-did-cities-actually-offer-amazon/559220/.

46. Sharon Zukin, *Naked City: The Death and Life of Authentic Urban Places* (Oxford, UK: Oxford University Press, 2011), 2.

47. Richard L. Florida, *The Rise of the Creative Class: And How It's Transforming Work, Leisure, Community and Everyday Life* (New York: Basic Books, 2002); Jamie Peck, "Struggling with the Creative Class," *International Journal of Urban and Regional Research* 29, no. 4 (2005): 740–70.

48. Meghan Ashlin Rich, "'From Coal to Cool': The Creative Class, Social Capital, and the Revitalization of Scranton," *Journal of Urban Affairs* 35, no. 3 (2013): 365–84; Jeffrey Zimmerman, "From Brew Town to Cool Town: Neoliberalism and the Creative City Development Strategy in Milwaukee," *Cities* 25, no. 4 (2008): 230–42.

49. Norma M. Rantisi and Deborah Leslie, "Materiality and Creative Production: The Case of the Mile End Neighborhood in Montréal," *Environment & Planning A: Economy & Space* 42, no. 12 (2010): 2824–41.

50. Tridib Banerjee and Anastasia Loukaitou-Sideris, "Competitions as a Design Method: An Inquiry," *Journal of Architectural and Planning Research* 7, no. 2 (1990): 116.

51. Gordon C. C. Douglas, *The Help-Yourself City: Legitimacy and Inequality in DIY Urbanism* (Oxford, UK: Oxford University Press, 2018).

52. City of New York, *Road Map for the Digital City: Achieving New York City's Digital Future* (New York: NYC Digital, 2011), http://www.nyc.gov/html/media/media/PDF/90dayreport.pdf.

53. Michael Cataldi et al., "Residues of a Dream World: The High Line, 2011," *Theory, Culture & Society* 28, nos. 7–8 (December 2011): 358–89; Steven Lang and Julia Rothenberg, "Neoliberal Urbanism, Public Space, and the Greening of the Growth Machine: New York City's High Line Park," *Environment & Planning A: Economy & Space* 49, no. 8 (August 2017): 1743–61; Nate Millington, "From Urban

Scar to 'Park in the Sky': Terrain Vague, Urban Design, and the Remaking of New York City's High Line Park," *Environment & Planning A: Economy & Space* 47, no. 11 (November 2015): 2324–38.

54. City of New York, *Road Map for the Digital City.*

55. NYC.gov, "Re-Invent Payphones Information Session."

56. Sherry R. Arnstein, "A Ladder of Citizen Participation," *Journal of the American Institute of Planners* 35, no. 4 (1969): 220.

57. Ajay Garde, "Citizen Participation, Design Competition and the Product in Urban Design: Insights from the Orange County Great Park," *Journal of Urban Design* 19, no. 1 (2014): 89–118.

58. Ute Lehrer, "Urban Design Competitions," in *Companion to Urban Design,* ed. Tridib Banerjee and Anastasia Loukaitou-Sideris (New York: Routledge, 2011), 304.

59. Charles V. Bagli, "Going Out with Building Boom, Mayor Pushes Billions in Projects," *New York Times,* December 15, 2013, https://www.nytimes.com/2013/12/16/nyregion/going-out-with-building-boom-mayor-pushes-billions-in-projects.html; C. J. Hughes, "Making Nice with Mike: Developers with Close Ties to City Hall, and Who's Out of Favor," *Real Deal New York,* April 1, 2012, https://therealdeal.com/issues_articles/making-nice-with-mike-developers-with-closes-ties-to-city-hall-and-whos-out-of-favor/; E. B. Solomont and Georgia Kromrei, "Love, Hate, and Real Estate: Bloomberg's Record May Polarize Voters," *Real Deal New York,* November 26, 2019, https://therealdeal.com/2019/11/26/love-hate-and-real-estate-bloombergs-record-may-polarize-voters/.

60. Oremus, "Michael Bloomberg Wants to Do Something Cool"; Tim Smedley, "Top-Down or Bottom-Up? Two Visions of Smart Cities," *New Scientist,* December 4, 2013, https://www.newscientist.com/article/mg22029465-000-top-down-or-bottom-up-two-visions-of-smart-cities/.

61. New York City Department of Information Technology and Telecommunication (hereafter NYC DoITT), "Reinvent Payphones Challenge / LinkNYC Recognized as 2015 Harvard Ash Center Bright Idea in Government" (press release), n.d., https://www1.nyc.gov/assets/doitt/downloads/pdf/Bright_Ideas_Press_Release_FiNAL.pdf.

62. NYC DoITT, "Request for Information Regarding the Future of Public Pay Telephones on New York City Sidewalks and Potential Alternative or Additional Forms of Telecommunications Facilities on New York City Sidewalks" (hereafter "RFi"), July 11, 2012, https://www1.nyc.gov/assets/doitt/downloads/pdf/payphone_rfi.pdf.

63. Out-of-home ad spending in New York City was $701.66 million in 2015; in Los Angeles, it was $412.97 million. See https://www.statista.com/statistics/593509/ooh-ad-markets-usa/.

64. Colin O'Donnell, David Rocamora, and Sholom Ellenberg, "Web-Level Engagement and Analytics for the Physical Space," US 0107732A1, filed May 7, 2012, issued May 2, 2013.

65. Gail Chiasson, "Titan Acquires All N.Y. Verizon Phone Kiosks," *Daily-DOOH,* August 20, 2010, http://www.dailydooh.com/archives/31780.

66. NYC DoITT, "Request for Information Regarding the Future of Public Pay Telephones on New York City Sidewalks and Potential Alternative or Additional Forms of Telecommunications Facilities on New York City Sidewalks: Response of Titan Outdoor Communications, Inc.," August 30, 2012, http://www.nyc.gov/ (specific page discontinued).

67. NYC DoITT, "Request for Proposals for a Franchise to Install, Operate and Maintain Public Communications Structures in the Boroughs of the Bronx, Brooklyn, Manhattan, Queens and Staten Island," 2014. https://www1.nyc.gov/assets/doitt/downloads/pdf/DoITT-Public-Communication-Structure-RFP-4-30-14.pdf; NYC DoITT, "RFi."

68. Stanley Shor, Assistant Commissioner, Franchise Administration, NYC DoITT, personal communication, July 28, 2017.

69. Byford, "New York City Launches Competition to Create the Perfect Payphone."

70. Joseph Flaherty, "10-Foot-Tall Double-Sided Touchscreen Wins NYC's Pay Phone Redesign Challenge," *Wired,* March 16, 2013, https://www.wired.com/2013/03/nyc-pay-phones-redesign-challenge/.

71. Lucy Suchman, Randall Trigg, and Jeanette Blomberg, "Working Artefacts: Ethnomethods of the Prototype," *British Journal of Sociology* 53, no. 2 (2002): 163.

72. Samuel Kinsley, "Futures in the Making: Practices to Anticipate 'Ubiquitous Computing,'" *Environment & Planning A: Economy & Space* 44, no. 7 (2012): 1565; see also Samuel Kinsley, "Representing 'Things to Come': Feeling the Visions of Future Technologies," *Environment & Planning A: Economy & Space* 42, no. 11 (2010): 2771–90; Samuel Kinsley, "Anticipating Ubiquitous Computing: Logics to Forecast Technological Futures," *Geoforum* 42, no. 2 (2011): 231–40.

73. Bo Begole and Ryusuke Masuoka, quoted in Kinsley "Anticipating Ubiquitous Computing," 237.

74. Kinsley, "Representing 'Things to Come,'" 8.

75. Kinsley, 7.

76. Pierre Bourdieu, *Language and Symbolic Power,* ed. John Thompson, trans. Gino Raymond and Matthew Adamson (Cambridge, Mass.: Harvard University Press, 1999).

77. Bourdieu.

78. Turi McKinley and Ethan Imboden, "The Evolution of a Smart City Hub," Frog Design, 2019, https://www.frogdesign.com/designmind/evolution-of-a-smart-city-hub.

79. Ben Green, *The Smart Enough City: Putting Technology in Its Place to Reclaim Our Urban Future* (Cambridge, Mass.: MIT Press, 2019).

80. Tanvi Misra, "De Blasio's Vision for New York: Broadband for All by 2025," *CityLab,* April 21, 2015, http://www.citylab.com/tech/2015/04/de-blasios-vision-for-new-york-broadband-for-all-by-2025/391092/.

81. NYC DoITT, "The Future of the Payphone," 2012, http://www.nyc.gov/ (specific page discontinued); NYC DoITT, "RFi."

82. Devin Coldewey, "Average Broadband Speed in U.S. Rises above 50 Megabits for the First Time," *TechCrunch,* August 3, 2016, http://social.techcrunch.com/2016/08/03/average-broadband-speed-in-us-rises-above-50-megabits-for-the-first-time/.

83. Gustin, "LinkNYC's New Free Network Is Blazing Fast."

84. Pinto, "Google Is Transforming NYC's Payphones."

85. Catherine Shu, "New York City Sues Verizon for Not Completing Citywide Fiber Network," *TechCrunch,* March 14, 2017, http://social.techcrunch.com/2017/03/14/nyc-sues-verizon/; Greta Byrum, personal communication, March 9, 2017.

86. Joe Marvill, "Telebeam's LinkNYC Lawsuit Details Released," *Queens Tribune,* December 18, 2014, http://queenstribune.com/ (site discontinued).

87. Gibson Dunn, "Objection to NYC Wi-Fi Payphone Contract with CityBridge," December 8, 2014, https://www.scribd.com/document/249805478/Objection-to-NYC-wi-fi-Payphone-contract-with-CityBridge. Specifically, the Telebeam letter alleges that "Robert Richardson, a former senior executive at the Control Group, left the Control Group at the end of July 2014 to take a senior technology policymaking post to 'help direct technology strategy from City Hall' at the same time that DoITT was deliberating on the responses to the RFP."

88. Dunn.

89. Dunn.

90. Greg B. Smith, "De Blasio's Wi-Fi Plan Gives Slower Service to Poorer Neighborhoods," *New York Daily News,* November 24, 2014, http://www.nydailynews.com/new-york/exclusive-de-blasio-wi-fi-plan-slower-poor-nabes-article-1.2021146.

91. T. C. Sottek, "New York City's Ambitious Free Wi-Fi Plan Sounds Great, Unless You Live in a Poor Neighborhood," *The Verge,* November 24, 2014, https://www.theverge.com/2014/11/24/7275567/nyc-public-wifi-is-rich.

92. New York City Council, Committee on Technology, NYCC Resolution 0838-2015: Condemning the Inequitable Contract between NYC and LinkNYC Which Will Provide Better Wi-Fi Access to Areas throughout the City, September 17, 2015, cosponsored by council members Fernando Cabrera, Rosie Mendez, Darlene Mealy, and Deborah L. Rose.

93. Patrick McGeehan, "New Yorkers Greet the Arrival of Wi-Fi Kiosks with Panic, Skepticism and Relief," *New York Times,* July 27, 2016, https://www.nytimes.com/2016/07/27/nyregion/link-nyc-wi-fi-kiosks.html.

94. Nathan Ingraham, "Google's Sidewalk Labs Is Taking Over the Plan to Blanket NYC with Free Wi-Fi," *The Verge,* June 23, 2015, http://www.theverge.com/2015/6/23/8834863/google-sidewalk-labs-linknyc-free-wifi.

95. Larry Page, "Sidewalk Labs," *Google+* (blog), June 10, 2015, https://plus.google.com/ (site discontinued); C. J. Hughes, "Daniel Doctoroff Takes His Business to Hudson Yards," *New York Times,* January 26, 2016, https://www.nytimes.com/2016/01/27/realestate/commercial/daniel-doctoroff-takes-his-business-to-hudson-yards.html.

96. "CityBridge to Launch LinkNYC."

97. Gustin, "LinkNYC's New Free Network Is Blazing Fast."

98. New York Civil Liberties Union, "City's Public Wi-Fi Raises Privacy Concerns," March 16, 2016, http://www.nyclu.org/news/citys-public-wi-fi-raises-privacy-concerns.

99. Mark Harris, "Inside Alphabet's Money-Spinning, Terrorist-Foiling, Gigabit Wi-Fi Kiosks," *Recode,* July 1, 2016, https://www.recode.net/2016/7/1/12072122/alphabet-sidewalk-labs-city-wifi-sidewalk-kiosks; Kofman, "Are New York's Free LinkNYC Internet Kiosks Tracking Your Movements?"; Claire Lampen, "Yes, LinkNYC Kiosks Are Giant Data-Harvesting Surveillance Cameras, Obviously," *Gothamist,* April 25, 2019, http://gothamist.com/2019/04/25/linknyc_kiosks_watching_you.php; Pinto, "Google Is Transforming NYC's Payphones"; Green, *Smart Enough City.*

100. Although companies like Apple have attempted to use randomizers to scramble MAC addresses to thwart attempts at tracking iPhones, there is evidence that system managers can still track nonusers by MAC address, even with relatively unsophisticated techniques. See Jeremy Martin et al., "A Study of MAC Address Randomization in Mobile Devices and When It Fails," *Proceedings on Privacy Enhancing Technologies* 4 (March 2017): 365–83, http://arxiv.org/abs/1703.02874.

101. Kofman, "Are New York's Free LinkNYC Internet Kiosks Tracking Your Movements?"

102. Joseph Turow, *The Daily You: How the New Advertising Industry Is Defining Your Identity and Your Worth* (New Haven, Conn.: Yale University Press, 2013); Joseph Turow, *The Aisles Have Eyes: How Retailers Track Your Shopping, Strip Your Privacy, and Define Your Power* (New Haven, Conn.: Yale University Press, 2017); Joseph Turow and Nora Draper, "Advertising's New Surveillance Ecosystem," in *Routledge Handbook of Surveillance Studies,* ed. Kirstie Ball, Kevin D. Haggerty, and David Lyon (Abingdon, UK: Routledge, 2012), 133–40; Harrison Smith, "Metrics, Locations, and Lift: Mobile Location Analytics and the Production of Second-Order Geodemographics," *Information, Communication & Society* 22, no. 8 (2019): 1044–61; Hal R. Varian, "Online Ad Auctions," *American Economic Review* 99, no. 2 (2009): 430–34; Garett Sloane, "WTF Is Dynamic Allocation?," *Digiday,* April 14, 2016, https://digiday.com/media/wtf-dynamic-allocation-google-ad-auctions/;

Jennifer Valentino-DeVries et al., "Your Apps Know Where You Were Last Night, and They're Not Keeping It Secret," *New York Times*, December 10, 2018, https://www.nytimes.com/interactive/2018/12/10/business/location-data-privacy-apps.html.

103. McGuigan, "Automating the Audience Commodity," 2.

104. Matthew Crain, "Financial Markets and Online Advertising: Reevaluating the Dotcom Investment Bubble," *Information, Communication & Society* 17, no. 3 (2014): 371–84; Sarah Myers West, "Data Capitalism: Redefining the Logics of Surveillance and Privacy," *Business & Society* 58, no. 1 (2019): 20–41.

105. Louise Story and Miguel Helft, "Google Buys an Online Ad Firm for $3.1 Billion," *New York Times*, April 14, 2007, https://www.nytimes.com/2007/04/14/technology/14deal.html; Dean Schmid, "The History of Programmatic Advertising: Everything You Need to Know," *Disruptor*, August 13, 2017, https://www.disruptordaily.com/the-history-of-programmatic-advertising-everything-you-need-to-know/.

106. Intersection. "Place Exchange | Real Programmatic Out of Home," n.d., http://placeexchange.com, emphasis added.

107. Packer, "Epistemology Not Ideology," 298; see also Sven Brodmerkel and Nicholas Carah, *Brand Machines: Sensory Media and Calculative Culture* (London: Palgrave Macmillan, 2016).

108. Packer, 298, emphasis added.

109. McGuigan rightly argues that computational thinking and algorithmic practice have a much longer history in the media industries than we typically credit. See McGuigan, "Automating the Audience Commodity."

110. John Durham Peters, *The Marvelous Clouds: Toward a Philosophy of Elemental Media* (Chicago: University of Chicago Press, 2015), 23.

111. Ken Auletta, "How the Math Men Overthrew the Mad Men," *New Yorker*, May 21, 2018, https://www.newyorker.com/news/annals-of-communications/how-the-math-men-overthrew-the-mad-men.

112. Jörg Müller, Alex Schlottmann, and Antonio Krüuger, "Self-Optimizing Digital Signage Advertising," in *Adjunct Proceedings of the 9th International Conference on Ubiquitous Computing* (Berlin: Springer-Verlag, 2007), http://www.joergmueller.info/pdf/Ubicomp07MuellerSelfOptimizing.pdf.

113. Joseph Bernstein, Jeremy Singer-Vine, and Sarah Ryley, "Exclusive: Hundreds of Devices Hidden inside New York City Phone Booths," *BuzzFeed*, October 6, 2014, http://www.buzzfeed.com/josephbernstein/exclusive-hundreds-of-devices-hidden-inside-new-york-city-ph.

114. Beacons emit a low-energy radio signal to Bluetooth-enabled smartphones, pushing content to users' devices. As *AdAge* reported in 2015, with beacons, "marketers are merging digital and physical worlds to create seamless, predictive, personalized, and delightful environments that increase sales and brand metrics"; see Charles Fulford, "Retail Is About to Be Reinvented, Driven by Digital

Technologies," *AdAge,* August 28, 2015, http://adage.com/article/digitalnext/retail-reinvented/300129/.

115. LinkNYC, "Factsheet," n.d., https://www.link.nyc/assets/downloads/LinkNYC-Fact-Sheet.pdf.

116. O'Donnell, Rocamora, and Ellenberg, "Web-Level Engagement."

117. O'Donnell.

118. eMarketer, "Listen In."

119. Kristina Monllos, "These Out-of-Home Avengers Ads Displayed Showtimes at the Closest Theater to You," *AdWeek,* May 1, 2018, http://www.adweek.com/brand-marketing/these-out-of-home-avengers-ads-displayed-showtimes-at-the-closest-theater-to-you/.

120. eMarketer, "Listen In."

121. Ryan Calo, "Digital Market Manipulation," *George Washington Law Review* 82, no. 4 (2014): 995–1051, https://papers.ssrn.com/abstract=2309703.

122. Turow and Couldry, "Media as Data Extraction," 416, emphasis added.

123. Dallas W. Smythe, "Communications: Blindspot of Western Marxism," *Canadian Journal of Political and Social Theory* 1, no. 3 (1977): 1–28; Dallas W. Smythe, "On the Audience Commodity and Its Work," in *Dependency Road: Communications, Capitalism, Consciousness, and Canada* (Norwood, N.J.: Ablex, 1981). For a more recent treatment of Smythe's concepts, see Lee McGuigan and Vincent Manzerolle, eds., *The Audience Commodity in a Digital Age: Revisiting a Critical Theory of Commercial Media* (New York: Peter Lang, 2014).

124. Oscar Gandy, *The Panoptic Sort: A Political Economy of Personal Information* (Boulder, Colo.: Westview Press, 1993); David Lyon, ed., *Surveillance as Social Sorting: Privacy, Risk and Automated Discrimination* (New York: Routledge, 2003).

125. Mark A. Thomas, "Covering Up LinkNYC Cameras? Do It with a Smile," *The Payphone Project* (blog), May 27, 2019, https://www.payphone-project.com/covering-up-linknyc-cameras-do-it-with-a-smile.html.

126. Rethink LinkNYC, "The Better to Spy You With," October 3, 2017, http://rethinklink.nyc/2017/10/03/the-better-to-spy-you-with.html.

127. One example is the Red Hook Initiative, a "self-healing mesh network" powered by solar panels. Community members lead decision-making on where servers and data are stored and have designed a platform for community organizing. See https://redhookwifi.org/about/mission/. Greta Byrum, personal communication, March 9, 2017.

128. Kendra Chamberlain, "Municipal Broadband Is Roadblocked or Outlawed in 26 States," *Broadband Now,* April 17, 2019, https://broadbandnow.com/report/municipal-broadband-roadblocks/.

129. Provenzano, "Why Google Needs to Focus on Connected Cities."

130. Steve Hamilton and Ximon Zhu, *Funding and Financing Smart Cities* (Washington, D.C.: Deloitte Center for Government Insights, 2017), https://www2

.deloitte.com/content/dam/Deloitte/us/Documents/public-sector/us-ps-funding-and-financing-smart-cities.pdf.

131. Linda Huber, "Is New York City's Public Wi-Fi Actually Connecting the Poor?" *Motherboard,* September 14, 2016, https://motherboard.vice.com/en_us/article/vv7pw3/linknyc-is-bringing-internet-to-new-yorks-most-disconnected-people.

132. NYC Office of the Mayor, "Mayor de Blasio Announces Public Launch of LinkNYC Program," February 18, 2016, http://www1.nyc.gov/office-of-the-mayor/news/184-16/mayor-de-blasio-public-launch-linknyc-program-largest-fastest-free-municipal.

133. Guillaume Latzko-Toth, Johan Söderberg, Florence Millerand, and Steve Jones, "Misuser Innovations: The Role of 'Misuses' and 'Misusers' in Digital Communication Technologies," in *DigitalSTS: A Field Guide for Science & Technology Studies,* ed. Janet Vertesi and David Ribes (Princeton, N.J.: Princeton University Press, 2019), 393–411.

134. McGeehan, "New Yorkers Greet."

135. McGeehan.

136. NYC Office of the Mayor, "Transcript: Mayor de Blasio Appears Live on WNYC," September 16, 2016, http://www1.nyc.gov/office-of-the-mayor/news/741-16/transcript-mayor-de-blasio-appears-live-wnyc.

137. For a critique of "order-maintenance policing," see Bernard E. Harcourt, *Illusion of Order: The False Promise of Broken Windows Policing* (Cambridge, Mass.: Harvard University Press, 2001). "Broken windows"–style policing has also had lingering effects on perceptions of everyday life in the city; see, for instance, Matthew M. Wilson, "Data Matter(s): Legitimacy, Coding, and Qualifications-of-Life," *Environment & Planning D: Society & Space* 29 (2011): 857–72.

138. Shannon Mattern, "Public In/Formation," *Places Journal,* November 2016, https://doi.org/10.22269/161115.

139. William H. Whyte, *The Social Life of Small Urban Spaces* (New York: Project for Public Spaces, 2001). For an example of this coverage, see Jamie Condliffe, "New York City Gave People Free Internet, and They Used It to Look at Porn," *MIT Technology Review,* September 15, 2016, https://www.technologyreview.com/s/602390/humans-do-dumb-things-with-smart-cities/.

140. Tim Cresswell, "Towards a Politics of Mobility," *Environment & Planning D: Society & Space* 28, no. 1 (2010): 17–31.

141. Cresswell, 21.

142. Laura Bliss, "If Google Were Mayor," *The Atlantic,* January 10, 2018, https://www.theatlantic.com/business/archive/2018/01/google-toronto-smart-city-quayside/550127/; Laura Bliss, "Behind the Backlash over Sidewalk Labs' Smart City," *CityLab,* September 7, 2018, https://www.citylab.com/design/2018/09/how-smart-should-a-city-be-toronto-is-finding-out/569116/.

143. Sidewalk Labs, "Responsive Architecture," *City of the Future Podcast,* n.d., https://medium.com/sidewalk-talk/city-of-the-future-podcast-episode-10-re sponsive-architecture-3914ae11a75b.

144. Bliss, "Behind the Backlash."

145. Draper and Turow, "Corporate Cultivation of Digital Resignation."

146. Unless specified otherwise, all quotations in this section are from Alex Levin and Meredith Popolo, personal communication, December 29, 2016.

147. Jenna Wortham, "How New Yorkers Adjusted to Sudden Smartphone Withdrawal," *New York Times,* November 3, 2012, https://bits.blogs.nytimes.com/2012/11/03/how-new-yorkers-adjusted-to-sudden-smartphone-withdrawal/.

148. "Hurricane Sandy Fast Facts," *CNN,* July 13, 2013, http://www.cnn.com/2013/07/13/world/americas/hurricane-sandy-fast-facts/index.html.

149. Alex Levin and Meredith Popolo, "NYC Via, Summary," 2013, http://nycvia.tumblr.com/?og=1.

150. Fiona Campbell, *Contours of Ableism: The Production of Disability and Abledness* (New York: Palgrave Macmillan, 2009).

151. Jennifer Valentino-DeVries, "Tracking Phones, Google Is a Dragnet for the Police," *New York Times,* April 13, 2019, https://www.nytimes.com/inter-active/2019/04/13/us/google-location-tracking-police.html; Jennifer Valentino-DeVries et al., "Your Apps Know Where You Were Last Night, and They're Not Keeping It Secret," *New York Times,* December 10, 2018, https://www.nytimes.com/interactive/2018/12/10/business/location-data-privacy-apps.html.

152. Unless specified otherwise, all quotations in this section are from Nick Wong and Ann Chen, personal communication, December 13, 2016.

153. Jon Brodkin, "Opinion: When the Landline Is a Lifeline," *New York Times,* June 5, 2014, https://www.nytimes.com/2014/06/05/opinion/when-the-landline-is-a-lifeline.html.

154. Ann Chen, "Windchimes," n.d., http://annhchen.com/Windchimes.

155. Unless specified otherwise, all quotations in this section are from William Arnold and Andrei Juradowitch, personal communication, January 10, 2017.

156. Shannon Mattern, *Code and Clay, Data and Dirt: Five Thousand Years of Urban Media* (Minneapolis: University of Minnesota Press, 2017).

157. William Arnold and Andrei Juradowitch, "Digital Democracy: A Voice for the People of NYC," 2013, http://ucop.com.au/nycphone/.

158. Johanna Drucker, "Species of Espaces and Other Spurious Concepts Addressed to Reading the Invisible Features of Signs within Systems of Relations," *Design and Culture* 2, no. 2 (2010): 135–53; Mattern, *Code and Clay.*

159. Celia Nowell, "A Comic Artist with Eyes on the City Is Now Seeing Himself All Over Town," *Bedford + Bowery,* February 28, 2019, https://bedfordandbowery.com/2019/02/a-comic-artist-with-eyes-on-the-city-is-now-seeing-himself-all-over-town/.

160. R. J. Rushmore, "NYC Kiosks Invite Artists to Pay Nearly $1,000 to Show Their Work," *Hyperallergic*, January 15, 2019, https://hyperallergic.com/479769/nyc-kiosks-invite-artists-to-pay-nearly-1000-to-show-their-work/.

161. Jonah Gaynor, "LinkNYC's Local Advertising Problem: Small Business Ads Not Wanted on Wi-Fi Kiosks," *NYC Tech News*, July 23, 2016, http://silicon.nyc/local-business-advertising-linknyc-wifi/.

162. Mattern, "Interfacing Urban Intelligence"; Parks, "Stuff You Can Kick."

163. "Peek-a-Boo, Google Sees You."

164. Provenzano argues that the fall in ad prices is the result of increased competition with Facebook and a greater share of Alphabet's ad sales coming from YouTube, which are cheaper because they are less targeted than banner and search ads. See Provenzano, "Why Google Needs to Focus on Connected Cities."

165. Provenzano.

166. Daniel Greene and Daniel Joseph, "The Digital Spatial Fix," *TripleC: Communication, Capitalism & Critique* 13, no. 2 (2015): 223–48; David Harvey, *The Limits to Capital* (London: Verso, 1982).

167. Provenzano, "Why Google Needs to Focus on Connected Cities."

168. Provenzano.

169. Hamilton and Zhu, *Funding and Financing Smart Cities*, 8.

170. Barry Frey, "DOOH Companies Positioned to Lead 'Smart Cities' Initiatives," *MediaVillage*, December 15, 2016, https://www.mediavillage.com/article/dooh-companies-positioned-to-lead-smart-cities-initiatives/.

171. Rahul Kumar and Supradip Baul, "Digital-Out-Of-Home (DOOH) Market by Format Type," *Allied Market Research*, February 2018, https://www.alliedmarketresearch.com/digital-out-of-home-DOOH-market.

172. Shelly Singh, "Digital Out of Home Market Worth $32.1 Billion by 2025 with a Growing CAGR of 10.7%," *Markets and Markets*, September 2017, https://www.marketsandmarkets.com/PressReleases/dooh.asp.

173. PricewaterhouseCoopers, in partnership with Interactive Advertising Bureau, *Growing Programmatic DOOH: Opportunities and Challenges* (London: PricewaterhouseCoopers, 2019), https://www.iab.com/wp-content/uploads/2019/09/IAB_2019-09-17_Growing-pDOOH.pdf.

174. Stuart Hall, "Encoding/Decoding," in *Media and Cultural Studies: Keyworks*, ed. Meenakshi Gigi Durham and Douglas M. Kellner (Malden, Mass.: Blackwell, 2001), 163–73; Packer, "Epistemology Not Ideology."

175. McGuigan notes that as early as the 1950s and '60s, the discourse of digitization in the advertising industry revolved around the capture of new efficiencies, with marketing firms working to build "logistical utilities to service the needs and ambitions of a media system organized around the twin goals of selling audiences and selling sponsors' products." McGuigan, "Automating the Audience Commodity," 17.

176. eMarketer, "Listen In."

177. Steven J. Jackson, "Rethinking Repair," in *Media Technologies: Essays on Communication, Materiality, and Society,* edited by Tarleton Gillespie, Pablo J. Boczkowski, and Kirsten A. Foot (Cambridge, Mass.: MIT Press, 2014), 221–40.

178. Anthony Dunne and Fiona Raby, *Speculative Everything: Design, Fiction and Social Dreaming* (Cambridge, Mass.: MIT Press, 2013), 35.

2. Control

1. Field notes, December 17, 2015.

2. Carlo Ratti and Matthew Cladel, "Life in the Uber City," *Design Indaba,* July 14, 2014, http://www.designindaba.com/articles/point-view/opinion-life-uber-city.

3. Jennifer Burns, *Goddess of the Market: Ayn Rand and the American Right* (Oxford, UK: Oxford University Press, 2009).

4. Mary L. Gray and Siddharth Suri, *Ghost Work: How to Stop Silicon Valley from Building a New Global Underclass* (Boston: Houghton Mifflin Harcourt, 2019).

5. Philip Dray, *There Is Power in a Union: The Epic Story of Labor in America* (Norwell, Mass.: Anchor, 2011); Sarah Kessler, *Gigged: The End of the Job and the Future of Work* (New York: St. Martin's Press, 2018); Erik Loomis, *A History of America in Ten Strikes* (New York: New Press, 2018).

6. Jessi Hempel, "Video and Transcript: Uber CEO Travis Kalanick," *Fortune,* July 23, 2013, https://fortune.com/2013/07/23/video-and-transcript-uber-ceo-travis-kalanick/.

7. Platform labor is increasingly gamified, with bonuses and competitions woven into the fabric of everyday work. See Casey O'Donnell, "Getting Played: Gamification and the Rise of Algorithmic Surveillance," *Surveillance & Society* 12, no. 3 (2014): 349–59; Sarah Mason, "High Score, Low Pay: Why the Gig Economy Loves Gamification," *The Guardian,* November 20, 2018, https://www.theguardian.com/business/2018/nov/20/high-score-low-pay-gamification-lyft-uber-drivers-ride-hailing-gig-economy; Jamie Woodcock and Mark R. Johnson, "Gamification: What It Is, and How to Fight It," *Sociological Review* 66, no. 3 (2018): 542–58; Trebor Scholz, *Uberworked and Underpaid: How Workers Are Disrupting the Digital Economy* (Malden, Mass.: Polity, 2016).

8. Ugo Rossi, "The Common-Seekers: Capturing and Reclaiming Value in the Platform Metropolis," *Environment & Planning C: Politics & Space* 37, no. 8 (2019): 1418–33; see also Miriam Greenberg and Penny Lewis, eds., *The City Is the Factory: New Solidarities and Spatial Strategies in an Urban Age* (Ithaca, N.Y: ILR Press, 2017).

9. Marco Briziarelli, "Spatial Politics in the Digital Realm: The Logistics/Precarity Dialectics and Deliveroo's Tertiary Space Struggles," *Cultural Studies* 33, no. 5 (2019): 823–40.

10. The "future of work" is a common characterization of the platform economy and always seems to foreclose alternative organizational paradigms. See, for

example, Deloitte's collection of reports on "Technology and the Future of Work," which identifies two parallel forces influencing labor futures: "The growing adoption of artificial intelligence in the workplace, and the expansion of the workforce to include both on- and off-balance-sheet talent;" available at https://www2.deloitte.com/us/en/insights/focus/technology-and-the-future-of-work.html.

11. Martin Kenney and John Zysman, "The Rise of the Platform Economy," *Issues in Science and Technology* (blog), March 29, 2016, https://issues.org/the-rise-of-the-platform-economy/.

12. Nick Srnicek, *Platform Capitalism* (Malden, Mass.: Polity, 2016).

13. Sarah Barns, *Platform Urbanism: Negotiating Platform Ecosystems in Connected Cities* (Singapore: Springer, 2020).

14. Jack Linchuan Qiu, Melissa Gregg, and Kate Crawford, "Circuits of Labour: A Labour Theory of the iPhone Era," *tripleC: Communication, Capitalism & Critique* 12, no. 2 (2014): 564–81.

15. There are exceptions to this. Instacart, for example, hires both employees and independent contractors, primarily because training independent contractors violates regulations; cf. Ellen Huet, "Instacart Makes Some Contractors Employees So It Can Train Them More," *Forbes,* June 22, 2015, https://www.forbes.com/sites/ellenhuet/2015/06/22/instacart-makes-some-contractors-employees-training/. See also Jing Dong and Rouba Ibrahim, "Flexible Workers or Full-Time Employees? On Staffing Systems with a Blended Workforce," *SSRN Scholarly Paper,* May 22, 2017, https://papers.ssrn.com/abstract=2971841.

16. Some 40 percent of on-demand drivers in New York City qualify for Medicaid, the public healthcare option in the United States only available to those at or below the poverty line, and another 18 percent qualify for food stamps; Alexia Fernández Campbell, "The Worldwide Uber Strike Is a Key Test for the Gig Economy," *Vox,* May 8, 2019, https://www.vox.com/2019/5/8/18535367/uber-drivers-strike-2019-cities.

17. Barbara Ehrenreich, *Nickel and Dimed: On (Not) Getting By in America* (New York: Picador, 2011); Debra Howcroft and Helen Richardson, eds., *Work and Life in the Global Economy: A Gendered Analysis of Service Work* (New York: Palgrave Macmillan, 2009); David K. Shipler, *The Working Poor: Invisible in America* (New York: Vintage, 2004).

18. Alessia Pisoni and Alberto Onetti, "When Startups Exit: Comparing Strategies in Europe and the USA," *Journal of Business Strategy* 39, no. 3 (2018): 26–33.

19. Srnicek, *Platform Capitalism.*

20. Srnicek; see also Nick Srnicek, "The Challenges of Platform Capitalism: Understanding the Logic of a New Business Model," *Juncture: Journal of the Institute for Public Policy Research* 23, no. 4 (2017), https://www.ippr.org/juncture-item/the-challenges-of-platform-capitalism.

21. We see these effects, for example, in New York City, where Uber's entrance to the ride-hailing market caused the price of a taxi medallion to drop to a mere eighth of its former cost, reflecting a devastating depreciation in the market's valuation of the yellow cabs; Brian M. Rosenthal, "As Thousands of Taxi Drivers Were Trapped in Loans, Top Officials Counted the Money," *New York Times,* May 19, 2019, https://www.nytimes.com/2019/05/19/nyregion/taxi-medallions.html.

22. Julia Tomassetti, "The Contracting/Producing Ambiguity and the Collapse of the Means/Ends Distinction in Employment," *South Carolina Law Review* 66 (2014): 315; Marina Lao, "Workers in the 'Gig' Economy: The Case for Extending the Antitrust Labor Exemption," *University of California, Davis Law Review* 51 (2017): 1543–87.

23. Quoted in Alex Press, "Code Red: Organizing the Tech Sector," *N+1,* April 18, 2018, https://nplusonemag.com/issue-31/politics/code-red/.

24. Press.

25. Srnicek, *Platform Capitalism,* 40; Srnicek, "Challenges of Platform Capitalism"; Adi Kamdar, "Why Some Gig Economy Startups Are Reclassifying Workers as Employees," *On Labor,* February 19, 2016, http://onlabor.org/why-some-gig-economy-startups-are-reclassifying-workers-as-employees/.

26. Gérard P. Cachon, Kaitlin M. Daniels, and Ruben Lobel, "The Role of Surge Pricing on a Service Platform with Self-Scheduling Capacity," *Manufacturing & Service Operations Management* 19, no. 3 (2017): 368–84; Itai Gurvich, Martin Lariviere, and Antonio Moreno, "Operations in the On-Demand Economy: Staffing Services with Self-Scheduling Capacity," *SSRN Scholarly Paper,* June 26, 2016, https://papers.ssrn.com/abstract=2336514; Terry Taylor, "On-Demand Service Platforms," *SSRN Scholarly Paper,* September 1, 2017, https://papers.ssrn.com/abstract=2722308.

27. Lily Irani, "Difference and Dependence among Digital Workers: The Case of Amazon Mechanical Turk," *South Atlantic Quarterly* 114, no. 1 (2015): 225–34; Karen E. C. Levy, "The Contexts of Control: Information, Power, and Truck-Driving Work," *Information Society* 31 (2015): 160–74; Alex Rosenblat and Luke Stark, "Algorithmic Labor and Information Asymmetries: A Case Study of Uber's Drivers," *International Journal of Communication* 10 (2016): 3758–84. On long-distance control as a source of power broadly, see John Law, "On the Methods of Long-Distance Control: Vessels, Navigation and the Portuguese Route to India," in *Power, Action and Belief: A New Sociology of Knowledge?* ed. John Law (London: Routledge, 1986), 234–63; Torin Monahan, "Dreams of Control at a Distance: Gender, Surveillance, and Social Control," *Cultural Studies ↔ Critical Methodologies* 9, no. 2 (2009): 286–305; Nikolas Rose and Peter Miller, "Political Power beyond the State: Problematics of Government," *British Journal of Sociology* 61, no. 1 (2010): 271–303.

28. Ryan Calo and Alex Rosenblat, "The Taking Economy: Uber, Information, and Power," *Columbia Law Review* 117, no. 6 (2017): 1623–90.

29. Julia Ticona, "Left to Our Own Devices: Navigating the Risks of Work and Love with Personal Technologies" (PhD. diss., University of Virginia, 2016), https://doi.org/10.18130/v3vg4d.

30. Geoff Nunberg, "Goodbye Jobs, Hello 'Gigs': How One Word Sums Up a New Economic Reality," *NPR*, January 11, 2016, http://www.npr.org/2016/01/11/460698077/goodbye-jobs-hello-gigs-nunbergs-word-of-the-year-sums-up-a-new-economic-reality.

31. u/schmuber, "What Oversaturation Looks Like," Reddit, June 20, 2018, https://www.reddit.com/r/uberdrivers/comments/8sl75w/what_oversaturation_looks_like/.

32. Robert Graboyes, "Gigs, Jobs, and Smart Machines," *Manhattan Institute: E21* (blog), April 26, 2016, https://economics21.org/html/gigs-jobs-and-smart-machines-1779.html.

33. Graboyes.

34. "TryCaviar.com—Courier" https://vimeo.com/126538031.

35. Valerio De Stefano, *The Rise of the "Just-in-Time Workforce": On-Demand Work, Crowd Work and Labour Protection in the Gig-Economy* (Geneva: International Labour Office, 2016), http://www.ilo.org/wcmsp5/groups/public/---ed_protect/---protrav/---travail/documents/publication/wcms_443267.pdf.

36. Ursula Huws, "Logged Labour: A New Paradigm of Work Organisation?" *Work Organisation, Labour & Globalisation* 10, no. 1 (2016): 7–26.

37. Wildgoose59, quoted in Novara Media, "The Gig Economy Is Killing People" (YouTube video, uploaded June 25, 2019), https://www.youtube.com/watch?v=rpth95YRgbQ.

38. Donald McNeill, "Global Firms and Smart Technologies: IBM and the Reduction of Cities," *Transactions of the Institute of British Geographers* 40, no. 4 (2015): 562–74; Deloitte, "Smarter Insights with Workforce Data Analytics," May 27, 2010, https://www2.deloitte.com/us/en/pages/deloitte-analytics/solutions/analytics-in-action-workforce.html; Matt Vella, "Meet the Future of Consulting," *Fortune*, December 21, 2011, https://fortune.com/2011/12/21/meet-the-future-of-consulting/; Robert Kuttner, "Labor Market Regulation and the Global Economic Crisis," in *Rethinking Workplace Regulation: Beyond the Standard Contract of Employment,* ed. Katherine V. W. Stone and Harry Arthurs (Cambridge, UK: Cambridge University Press, 2013), 42–57.

39. Huws, "Logged Labour."

40. Jeremias Prassl, *Humans as a Service: The Promise and Perils of Work in the Gig Economy* (Oxford, UK: Oxford University Press, 2018); see also Lilly Irani, "The Cultural Work of Microwork," *New Media & Society* 17, no. 5 (2015): 720–39; Irani, "Difference and Dependence among Digital Workers."

41. Lukas Biewald, quoted in Moshe Z. Marvit, "How Crowdworkers Became the Ghosts in the Digital Machine," *The Nation,* February 5, 2014, https://www.thenation.com/article/how-crowdworkers-became-ghosts-digital-machine/.

42. Geoff Cutmore, "Interview with Travis Kalanick, CEO and Co-Founder of Uber," *CNBC,* March 28, 2016, http://www.cnbc.com/2016/03/28/cnbc-transcript-interview-with-travis-kalanick-ceo-and-co-founder-of-uber.html.

43. Andrew Sayer, "New Developments in Manufacturing: The Just-in-Time System," *Capital & Class* 10, no. 3 (1986): 43–72; Graham Sewell and Barry Wilkinson, "'Someone to Watch Over Me': Surveillance, Discipline and the Just-in-Time Labour Process," *Sociology* 26, no. 2 (1992): 271–89.

44. Huws, "Logged Labour," 18.

45. Sayer, "New Developments in Manufacturing."

46. Sayer, 53.

47. Deborah Cowen, *The Deadly Life of Logistics: Mapping Violence in Global Trade* (Minneapolis: University of Minnesota Press, 2014); De Stefano, *Rise of the "Just-in-Time Workforce."*

48. Kathryn A. Cady, "Flexible Labor," *Feminist Media Studies* 13, no. 3 (2013): 395–414; Guy Standing, "Global Feminization through Flexible Labor," *World Development* 17, no. 7 (1989): 1077–95; Jose Antonio Gonzalez Anaya, *Labor Market Flexibility in Thirteen Latin American Countries and the United States : Revisiting and Expanding Okun Coefficients* (Washington, D.C.: World Bank, 1999), http://documents.worldbank.org/curated/en/864361468768724600/Labor-market-flexibility-in-thirteen-Latin-American-countries-and-the-United-States-revisiting-and-expanding-Okun-coefficients.

49. Alex Rosenblat, *Uberland: How Algorithms Are Rewriting the Rules of Work* (Berkeley: University of California Press, 2018).

50. Andrew Ross, *Nice Work If You Can Get It: Life and Labor in Precarious Times (New York: New York University Press, 2009)*; Srnicek, *Platform Capitalism.*

51. Jonathan V. Hall and Alan B. Krueger, "An Analysis of the Labor Market for Uber's Driver-Partners in the United States," *ILR Review* 71, no. 3 (2018): 705–32.

52. Brett Collins et al., "Is Gig Work Replacing Traditional Employment? Evidence from Two Decades of Tax Returns" (working paper), Department of Economics, University of Illinois at Urbana–Champaign, March 25, 2019, https://experts.illinois.edu/en/publications/is-gig-work-replacing-traditional-employment-evidence-from-two-de.

53. Danny Dorling, *Inequality and the 1%* (New York: Verso, 2014); Danny Dorling, *Injustice: Why Social Inequality Still Persists* (Bristol, UK: Policy Press, 2015); Chad Stone et al., "A Guide to Statistics on Historical Trends in Income Inequality," *Center on Budget and Policy Priorities,* November 28, 2011, https://www.cbpp.org/research/poverty-and-inequality/a-guide-to-statistics-on-historical-trends-in-income-inequality.

54. Rossi, "Common-Seekers," 1423.

55. Lizzie Richardson, "Platforms, Markets, and Contingent Calculation: The Flexible Arrangement of the Delivered Meal," *Antipode,* 2019, https://doi.org/10.1111/anti.12546.

56. Rossi, "Common-Seekers"; see also Calo and Rosenblat, "Taking Economy."

57. Dough, "Uber Vehicle Requirements: Can You Drive For Uber With Your Car?" *Ridesharing Driver* (blog), January 8, 2018, https://www.ridesharingdriver.com/uber-vehicle-requirements-can-you-drive-for-uber-with-your-car/. Taxi companies are no less exploitative. When drivers don't own their own car, they must rent it from the company, which takes a large chunk out of their daily earnings; see Biju Mathew, *Taxi! Cabs and Capitalism in New York City* (Ithaca, N.Y.: ILR Press, 2008).

58. Ngai Keung Chan, "The Rating Game: The Discipline of Uber's User-Generated Ratings," *Surveillance & Society* 17, nos. 1–2 (2019): 183–90; Rosenblat and Stark, "Algorithmic Labor and Information Asymmetries."

59. Vahine Vara, "Uber, Lyft, and Liability," *New Yorker,* November 4, 2014, https://www.newyorker.com/business/currency/uber-lyft-liability; Tim Fernholz, "A Coverage Dispute between Uber, Lyft, and Insurers Leaves Drivers Exposed," *Quartz,* March 23, 2015, https://qz.com/365854/a-coverage-dispute-between-uber-lyft-and-insurers-leaves-drivers-exposed/.

60. Lyft, "Insurance," 2018, https://help.lyft.com/hc/en-us/articles/115013080548-Insurance.

61. Juliana Feliciano Reyes, "Caviar Delivery Service to Offer Free Accident Insurance: Here's What That Means for Couriers," *Philadelphia Inquirer,* July 26, 2018, https://www.inquirer.com/philly/news/caviar-accident-insurance-couriers-pablo-avendano-onebeacon-20180726.html; Thomas Fox Parry, "The Death of a Gig Worker," *The Atlantic,* June 1, 2018, https://www.theatlantic.com/technology/archive/2018/06/gig-economy-death/561302/; George Ciccariello-Maher, "My Best Friend Lost His Life to the Gig Economy," *The Nation,* July 10, 2018, https://www.thenation.com/article/best-friend-lost-life-gig-economy/.

62. Novara Media, "Gig Economy Is Killing People."

63. Postmates "Terms of Service," updated December 19, 2019, https://about.postmates.com/legal/terms.

64. For example, see Bryan Menegus, "Workers Came Up with a Brilliant Plan to Use Postmates' Exploitative Platform against Itself," *Gizmodo,* May 31, 2019, https://gizmodo.com/workers-came-up-with-a-brilliant-plan-to-use-postmates-1835158791.

65. Ron Tal, "Is Uber Really Not a Taxi Service?" Quora, June 3, 2015, https://www.quora.com/Is-Uber-really-not-a-taxi-service (emphasis added).

66. Most often, misclassifications suits result in out-of-court settlements, largely because, as "atomized, dispersed actors," platform workers face a collective action problem that hampers their ability to organize. Consequently, surrogates (i.e., plaintiffs' attorneys) need to bring suit on the behalf of workers. However, these surrogates introduce conflicting interests with their bias toward large out-of-court settlements, which has deferred ruling on the classification question and allowed the platforms' rhetoric to generally set the tone for public

discourse. See Ruth Berins Collier, Veena B. Dubal, and Christopher L. Carter, "Disrupting Regulation, Regulating Disruption: The Politics of Uber in the United States," *Perspectives on Politics* 16, no. 4 (2018): 919–37; Veena B. Dubal, "Winning the Battle, Losing the War? Assessing the Impact of Misclassification Litigation on Workers in the Gig Economy," *Wisconsin Law Review* 2017, no. 4 (2017): 739–802.

67. Alison Griswold, "Uber's Secret Weapon Is Its Team of Economists," *Quartz*, October 14, 2018, https://qz.com/1367800/ubernomics-is-ubers-semi-secret-internal-economics-department/.

68. Griswold.

69. Orly Lobel, "Coase and the Platform Economy" (research paper, University of San Diego School of Law, 2017), http://dx.doi.org/10.2139/ssrn.3083764.

70. A 2016 conference dedicated to the on-demand economy, hosted by the MIT Sloan School of Management, opened with a discussion of how technologies reduce transaction costs. According to Eric Brynjolfsson and Andrew McAfee, directors of the MIT Initiative on the Digital Economy, a whole slew of activities that were previously labor intensive can now be accomplished with much greater ease thanks to new information and communications technologies. The example they offered was booking a house in Timbuktu.

71. Jessica Kim Cohen, "Lyft, Uber Expand Reach into Healthcare," *Modern Healthcare*, October 28, 2019, https://www.modernhealthcare.com/patients/lyft-uber-expand-reach-healthcare; Nanette Byrnes, "Uber for Education," *MIT Technology Review*, July 27, 2015, https://www.technologyreview.com/s/539106/uber-for-education/. See also Nick Srnicek, "Platform Monopolies and the Political Economy of AI," in *Economics for the Many*, ed. John McDonnell (New York: Verso, 2018), 152–63.

72. Ronald Coase, "The Nature of the Firm," *Economica* 4, no. 16 (1937): 388.

73. Joanne Yates, *Control through Communication: The Rise of System in American Management* (Baltimore: Johns Hopkins University Press, 1993).

74. Alfred D. Chandler, *The Visible Hand: The Managerial Revolution in American Business* (Cambridge, Mass.: Harvard University Press, 1977).

75. Wolfgang Schivelbusch, *The Railway Journey: The Industrialization of Time and Space in the Nineteenth Century* (Berkeley: University of California Press, 2014).

76. Richard N. Langlois, *The Dynamics of Industrial Capitalism: Schumpeter, Chandler, and the New Economy* (New York: Routledge, 2007); Richard Langlois, "The Vanishing Hand: The Changing Dynamics of Industrial Capitalism" (working paper, Department of Economics, University of Connecticut, 2002), http://digitalcommons.uconn.edu/econ_wpapers/200221; Richard Langlois, "Vertical (Dis)Integration and Technological Change," *Organizations and Markets* (blog), May 18, 2012, https://organizationsandmarkets.com/2012/05/18/vertical-disintegration-and-technological-change/; Christian Lechner, Gianni Lorenzoni, and

Enrico Tundis, "Vertical Disintegration of Production and the Rise of Market for Brands," *Journal of Business Venturing Insights* 6 (2016): https://doi.org/10.1016/j.jbvi.2016.05.002.

77. Langlois, "Vanishing Hand," 42.

78. Paul T. Kidd, *Agile Manufacturing: Forging New Frontiers* (Boston: Addison Wesley Longman, 1995); W. B. Lee and H. C. W. Lau, "Factory on Demand: The Shaping of an Agile Production Network," *International Journal of Agile Management Systems* 1, no. 2 (1999): 83–87; Claude R. Duguay, Sylvain Landry, and Federico Pasin, "From Mass Production to Flexible/Agile Production," *International Journal of Operations & Production Management* 17, no. 12 (1997): 1183–95; Yahaya Y. Yusuf, Mansoor Sarhadi, and Angappa Gunasekaran, "Agile Manufacturing: The Drivers, Concepts and Attributes," *International Journal of Production Economics* 62, nos. 1–2 (1999): 33–43.

79. Qiu, Gregg, and Crawford, "Circuits of Labour."

80. "There's an App for That," *The Economist,* December 30, 2014, https://www.economist.com/briefing/2014/12/30/theres-an-app-for-that.

81. Esko Kilpi, "The Future of Firms: Is There an App for That?" *Medium* (blog), February 16, 2015, https://medium.com/@EskoKilpi/movement-of-though-that-led-to-airbnb-and-uber-9d4da5e3da3a.

82. Julia Ballard, "New Orleans Startup TrayAway Celebrates 100th Location," *Silicon Bayou News,* January 16, 2020, http://siliconbayounews.com/2020/01/16/new-orleans-startup-trayaway-celebrates-100th-location/.

83. See also Anders Hansen Henten and Iwona Maria Windekilde, "Transaction Costs and the Sharing Economy," *INFO* 18, no. 1 (2016): 1–15; Orly Lobel, "Coase and the Platform Economy," *SSRN Scholarly Paper* (Rochester, NY: Social Science Research Network, 2018), https://papers.ssrn.com/abstract=3083764.

84. Julia Tomassetti, "Does Uber Redefine the Firm? The Postindustrial Corporation and Advanced Information Technology," *Hofstra Labor & Employment Law Journal* 34, no. 1 (2016): 1–78.

85. Tomassetti, 77.

86. Cowen, *Deadly Life of Logistics*; Jesse LeCavalier, *The Rule of Logistics: Walmart and the Architecture of Fulfillment* (Minneapolis: University of Minnesota Press, 2016); Ned Rossiter, *Software, Infrastructure, Labor: A Media Theory of Logistical Nightmares* (New York: Routledge, 2016).

87. Sewell and Wilkinson, "Someone to Watch Over Me," 271.

88. Anja Kanngieser, "Tracking and Tracing: Geographies of Logistical Governance and Labouring Bodies," *Environment & Planning D: Society & Space* 31, no. 4 (2013): 594–610; Ned Rossiter, "Locative Media as Logistical Media: Situating Infrastructure and the Governance of Labor in Supply-Chain Capitalism," in *Locative Media,* ed. Rowan Wilken and Gerard Goggin (New York: Routledge, 2014), 208–21.

89. Rossiter, "Locative Media as Logistical Media"; LeCavalier, *Rule of Logistics*; Emily Guendelsberger, *On the Clock: What Low-Wage Work Did to Me and How It Drives America Insane* (New York: Little, Brown, 2019).

90. Kanngieser, "Tracking and Tracing."

91. Manufacturing firms first integrated tracking systems into global supply chain networks at the behest of large, multinational companies like Walmart, which wield monopsonistic power over their suppliers; see LeCavalier, *Rule of Logistics*. But tracking is now standard across the logistics industries; see Tom Collins, "How Real-Time Information Can Transform Your Supply Chain," *Inbound Logistics,* June 21, 2016, http://www.inboundlogistics.com/cms/article/-how-real-time-info-can-transform-your-supply-chain/; Joseph O'Reilly, "Channeling the Brick and Click Dilemma," *Inbound Logistics,* August 21, 2015, http://www.inboundlogistics.com/cms/article/channeling-the-brick-and-click-dilemma/; Bill Powell, "What Is Your Fleet Trying to Tell You?" *Inbound Logistics,* March 3, 2016, http://www.inboundlogistics.com/cms/article/what-is-your-fleet-telling-you/.

92. Karen Levy and Solon Barocas, "Refractive Surveillance: Monitoring Customers to Manage Workers," *International Journal of Communication* 11 (2017): 1–24.

93. Brett Neilson, "Five Theses on Understanding Logistics as Power," *Distinktion: Journal of Social Theory* 13, no. 3 (2012): 330; Jennifer Valentino-DeVries et al., "Your Apps Know Where You Were Last Night, and They're Not Keeping It Secret" *New York Times,* December 10, 2018, https://www.nytimes.com/interactive/2018/12/10/business/location-data-privacy-apps.html.

94. Rossiter, "Locative Media as Logistical Media"; Rossiter, *Software, Infrastructure, Labor*; see also Ned Rossiter, "Coded Vanilla: Logistical Media and the Determination of Action," *South Atlantic Quarterly* 114, no. 1 (2015): 135–52.

95. Specifically, I interviewed thirteen couriers with Caviar and five on Postmates, and three drivers currently on Lyft and three currently on Uber (although all the drivers indicated that they have or continue to work for both platforms).

96. Sewell and Wilkinson, "Someone to Watch Over Me," 271.

97. Personal communication, August 24, 2015.

98. The names of the workers interviewed here have been changed to protect their identities.

99. Interview conducted March 2, 2016.

100. Moritz Altenried, "On the Last Mile: Logistical Urbanism and the Transformation of Labour," *Work Organisation, Labour & Globalisation* 13, no. 1 (2019): 114–29.

101. Interview conducted January 16, 2016.

102. Interview conducted March 7, 2016.

103. Interview conducted December 14, 2015.

104. Interview conducted February 16, 2016.

105. Interview conducted January 14, 2016.

106. Juliana Feliciano Reyes and Caitlin McCabe, "#OccupyPHA: Why People Have Been Protesting Outside of PHA Headquarters for 15 Days and Counting," *Philadelphia Inquirer,* May 8, 2019, https://www.inquirer.com/news/pha-philadelphia-sharswood-development-norman-blumberg-towers-20190508.html.

107. Christine Speer Lejeune, "Being Rich in Philly Sure Doesn't Look Like It Used To," *Philadelphia Magazine,* June 2, 2018, https://www.phillymag.com/news/2018/06/02/new-rich-people-philadelphia/; Danya Henninger, "Center City vs. the Neighborhoods: New Maps Show Philly's Stark Income Divide," *Billy Penn,* March 1, 2018, https://billypenn.com/2018/03/01/center-city-vs-the-neighborhoods-new-map-shows-phillys-stark-income-divide/; Jim Saksa, "Four Reasons Why Philadelphia Is Gentrifying," *PlanPhilly,* September 14, 2017, http://planphilly.com/articles/2017/09/14/four-reasons-why-philadelphia-is-gentrifying.

108. Interview conducted February 13, 2016.

109. Interview conducted February 13, 2016.

110. Interview conducted January 19, 2016.

111. Interview conducted March 3, 2016.

112. Noam Scheiber, "How Uber Drivers Decide How Long to Work," *New York Times,* September 5, 2016, https://www.nytimes.com/2016/09/05/business/economy/how-uber-drivers-decide-how-long-to-work.html; Noam Scheiber, "How Uber Uses Psychological Tricks to Push Its Drivers' Buttons," *New York Times,* April 2, 2017, https://www.nytimes.com/interactive/2017/04/02/technology/uber-drivers-psychological-tricks.html.

113. Interview conducted February 12, 2016.

114. Interview conducted February 16, 2016.

115. Juliana Feliciano Reyes and Jason Laughlin, "How Incentives in the Gig Economy Put Workers at Risk," *Philadelphia Inquirer,* May 31, 2018, https://www.inquirer.com/philly/news/gig-economy-worker-safety-independent-contractor-caviar-grubhub-handy-care-uber-20180531.html.

116. Sunil Rajaraman, "The On-Demand Economy Is a Bubble—and It's about to Burst," *Quartz,* April 28, 2017, https://qz.com/967474/the-on-demand-economy-is-a-bubble-and-its-about-to-burst/.

117. Alison Griswold, "Postmates Has Failed to Make Uber-for-Anything Cheap, and Is Quietly Misleading Customers about It," *Quartz,* August 15, 2016, https://qz.com/722230/postmates-has-failed-to-make-uber-for-anything-cheap/; Len Sherman, "Why Can't Uber Make Money?," *Forbes,* December 14, 2017, https://www.forbes.com/sites/lensherman/2017/12/14/why-cant-uber-make-money/#203b091410ec; Alex Wilhelm, "Understanding Why Lyft Loses Money," *Crunchbase News,* March 27, 2019, https://news.crunchbase.com/news/understanding-why-lyft-loses-money/.

118. Andrew J. Hawkins, "Uber Lost over $5 Billion in One Quarter, but Don't Worry, It Gets Worse," *The Verge,* August 8, 2019, https://www.theverge

.com/2019/8/8/20793793/uber-5-billion-quarter-loss-profit-lyft-traffic-2019; Conor Sen, "Food Delivery Looks Like Another Gig-Economy Dead End," *Bloomberg News,* October 31, 2019, https://www.bloomberg.com/opinion/articles/2019-10-31/food-delivery-is-a-dead-end-for-grubhub-doordash-and-postmates; Ranjan Roy, "Doordash and Pizza Arbitrage: There Is Such a Thing as Free Lunch," *The Margins,* May 17, 2020, https://themargins.substack.com/p/doordash-and-pizza-arbitrage.

119. Derek Thompson, "The Booming, Ethically Dubious Business of Food Delivery," *The Atlantic,* August 2, 2019, https://www.theatlantic.com/ideas/archive/2019/08/all-food-will-be-delivered/595222/.

120. Antonio Aloisi, "Commoditized Workers: Case Study Research on Labour Law Issues Arising from a Set of 'On-Demand / Gig Economy' Platforms," *Comparative Labor Law & Policy Journal* 37, no. 3 (2016): 653–90; De Stefano, *Rise of the "Just-in-Time Workforce."*

121. Kafui Attoh, Katie Wells, and Declan Cullen, "'We're Building Their Data': Labor, Alienation, and Idiocy in the Smart City," *Environment & Planning D: Society & Space* 37, no. 6 (2019): 1007–24; Declan Cullen, Kafui Attoh, and Kathryn Wells, "Taking Back the Wheel," *Dissent,* August 31, 2018, https://www.dissentmagazine.org/online_articles/uber-flying-car-silicon-valley-labor-technology-future-politics; Therese Poletti, "Uber and Lyft IPOs Mean the Cheap Rides Are Coming to an End," *MarketWatch,* May 13, 2019, https://www.marketwatch.com/story/uber-and-lyft-ipos-mean-the-cheap-rides-are-coming-to-an-end-2019-05-09.

122. Elizabeth Dunn, "How Delivery Apps May Put Your Favorite Restaurant Out of Business," *New Yorker,* February 3, 2018, https://www.newyorker.com/culture/annals-of-gastronomy/are-delivery-apps-killing-restaurants; Michael Sainato, "Uber and Lyft Drivers Say Apps are Short-Changing Wages While Raising Fares," *The Guardian,* April 18, 2019, https://www.theguardian.com/technology/2019/apr/18/uber-lyft-drivers-surge-pricing-wages; Megan Rose Dickey, "Postmates Workers Want Minimum Delivery Guarantees and at Least $15 per Hour," *TechCrunch,* May 20, 2019, http://social.techcrunch.com/2019/05/20/postmates-workers-want-minimum-delivery-guarantees-and-at-least-15-per-hour/.

123. Andrew Feenberg, *Between Reason and Experience: Essays in Technology and Modernity* (Cambridge, Mass.: MIT Press, 2010); Max Horkheimer, *Critique of Instrumental Reason,* trans. Matthew O'Connell (New York: Verso, 2013); Herbert Marcuse, *One-Dimensional Man: Studies in the Ideology of Advanced Industrial Society* (Boston: Beacon Press, 2012); Max Weber, *The Protestant Ethic and the Spirit of Capitalism,* trans. Talcott Parsons and Anthony Giddens (Boston: Unwin Hyman, 1930); Neil Fligstein, "Markets as Politics: A Political-Cultural Approach to Market Institutions," *American Sociological Review* 61, no. 4 (1996): 656–73.

124. Michel Callon and John Law, "On Qualculation, Agency, and Otherness," *Environment & Planning D: Society & Space* 23, no. 5 (2005): 718; Donald A.

MacKenzie, *An Engine, Not a Camera: How Financial Models Shape Markets* (Cambridge, Mass.: MIT Press, 2006); Donald MacKenzie, Fabian Muniesa, and Lucia Siu, eds., *Do Economists Make Markets? On the Performativity of Economics* (Princeton, N.J.: Princeton University Press, 2007); Michel Callon, Yuval Millo, and Fabian Muniesa, eds., *Market Devices* (Oxford, UK: Blackwell, 2007); Caitlin Zaloom, *Out of the Pits: Traders and Technology from Chicago to London* (Chicago: University of Chicago Press, 2006).

125. Michael Burawoy, *Manufacturing Consent: Changes in the Labor Process Under Monopoly Capitalism* (Chicago: University of Chicago Press, 1982).

126. Michel Callon and Fabian Muniesa, "Peripheral Vision: Economic Markets as Calculative Collective Devices," *Organization Studies* 26, no. 8 (2005): 1229–50.

127. See, for example, Sarah McBride, "Uber's Fight of California Data-Sharing Rule Highlights Its Bumpy Road," *Reuters*, December 19, 2014, https://www.reuters.com/article/us-uber-california-data-idUSKBN0JX01320141219; Casey Thomas, "Ride Oversharing: Privacy Regulation within the Gig Economy," *Cardozo Arts & Entertainment Law Journal* 36, no. 1 (2018): 247–76. See also Calo and Rosenblat, "Taking Economy," 54.

128. Thomas Claburn, "Uber Drivers Game Uber's System like Uber Games the Entire Planet," *Register UK*, August 2, 2017, https://www.theregister.co.uk/2017/08/02/uber_drivers_defy_machine_management/; Mason, "High Score, Low Pay."

129. Bernard W. Taylor, *Introduction to Management Science*, 11th ed. (Boston: Prentice Hall, 2012), 2.

130. Gurvich, Lariviere, and Moreno, "Operations in the On-Demand Economy."

131. Cachon, Daniels, and Lobel, "Role of Surge Pricing on a Service Platform," 369–70.

132. Michael Sheldon, "Income Targeting and the Ridesharing Market" (working paper, 2016), http://www.michaelsheldon.org/working-papers.

133. "How to Increase Your Earnings in Slow Periods Driving for Uber And Lyft," *Ridesharing Forum*, October 15, 2017, https://www.ridesharingforum.com/t/how-to-increase-your-earnings-in-slow-periods-driving-for-uber-and-lyft/60; Ridester, "The Best Time to Drive for Uber & Lyft: Monday & Friday Mornings," n.d., https://www.ridester.com/training/lessons/monday-friday-mornings/.

134. Ellen Huet, "Uber Data Show How Wildly Driver Pay Can Vary," *Forbes*, December 1, 2014, https://www.forbes.com/sites/ellenhuet/2014/12/01/uber-data-show-how-wildly-driver-pay-can-vary/.

135. Colin Camerer et al., "Labor Supply of New York City Cabdrivers: One Day at a Time," *Quarterly Journal of Economics* 112, no. 2 (1997): 407–41.

136. M. Keith Chen and Michael Sheldon, "Dynamic Pricing in a Labor Market: Surge Pricing and Flexible Work on the Uber Platform" (working paper, Uber / University of California, Los Angeles, December 11, 2015), http://citeseerx.ist.psu.edu/viewdoc/download?doi=10.1.1.704.3600&rep=rep1&type=pdf.

137. Rosenblat, *Uberland*, 133.

138. Rosenblat and Stark, "Algorithmic Labor and Information Asymmetries."

139. Interview conducted February 28, 2016.

140. Harish Guda and Upender Subramanian, "Your Uber Is Arriving: Managing On-Demand Workers through Surge Pricing, Forecast Communication, and Worker Incentives," *Management Science* 65, no. 5 (2019): 1949–2443; Harish Guda and Upender Subramanian, "Strategic Pricing and Forecast Communication on On-Demand Service Platforms" (working paper, October 11, 2017), https://pdfs.semanticscholar.org/.

141. Guda and Subramanian, "Strategic Surge Pricing."

142. Guda and Subramanian, "Your Uber Is Arriving," 3, emphasis added.

143. Cesar Conda and Derek Khanna, "Uber for Welfare," *Politico*, January 27, 2016, https://www.politico.com/agenda/story/2016/1/uber-welfare-sharing-gig-economy-000031.

144. Hempel, "Video and Transcript: Uber CEO Travis Kalanick," emphasis added.

145. Neilson, "Five Theses on Understanding Logistics as Power," 324.

146. Rossi, "Common-Seekers," 1423.

147. De Stefano, *Rise of the "Just-in-Time Workforce"*; see also Dennis Arnold and Joseph R. Bongiovi, "Precarious, Informalizing, and Flexible Work: Transforming Concepts and Understandings," *American Behavioral Scientist* 57, no. 3 (2013): 289–308.

148. Sandro Mezzadra and Brett Neilson, "On the Multiple Frontiers of Extraction: Excavating Contemporary Capitalism," *Cultural Studies* 31, nos. 2–3 (May 2017): 185–204; Rossi, "Common-Seekers."

149. Neilson, "Five Theses on Understanding Logistics as Power," 324; Karl Polanyi, *The Great Transformation: The Political and Economic Origins of Our Time* (Boston: Beacon Press, 2001); Joseph Schumpeter, *Capitalism, Socialism and Democracy* (London: Routledge, 1976).

150. Lora Kolodny, "Flush with Funding, Instacart Accelerates U.S. Expansion," *TechCrunch*, April 24, 2017, http://social.techcrunch.com/2017/04/24/flush-with-funding-instacart-accelerates-us-expansion/.

151. Chelsea Wahl, "Racing in the Aisles: How Algorithmic Management Structures Labor in On-Demand Grocery Work" (Ph.D. diss., University of Pennsylvania, in progress).

152. Guendelsberger, *On the Clock*; see also Kanngieser, "Tracking and Tracing."

153. Manuel Castells, *The Urban Question: A Marxist Approach* (Cambridge, Mass.: MIT Press, 1979); see also R. E. Pahl, "Castells and Collective Consumption," *Sociology* 12, no. 2 (1978): 309–15. On the city as a space of capital accumulation, see especially Altenried, "On the Last Mile"; Briziarelli, "Spatial Politics in the Digital Realm"; Rossi, "Common-Seekers."

154. Rossi, "Common-Seekers," 1423–24.

155. Levy, "Contexts of Control."

156. Altenried, "On the Last Mile," 118.

157. Martin Danyluk, "Capital's Logistical Fix: Accumulation, Globalization, and the Survival of Capitalism," *Environment & Planning D: Society & Space*, 36, no. 4 (2018): 630–47; see also Briziarelli, "Spatial Politics in the Digital Realm."

158. Marion Fourcade and Kieran Healy, "Seeing Like a Market," *Socio-Economic Review* 15, no. 1 (2017): 9–29; Graham Murdock, "Political Economies as Moral Economies: Commodities, Gifts, and Public Goods," in *The Handbook of Political Economy of Communications*, ed. Janet Wasko, Graham Murdock, and Helena Sousa (Malden, Mass.: Wiley-Blackwell, 2011), 11–40; E. P. Thompson, "The Moral Economy of the English Crowd in the Eighteenth Century," *Past & Present* 50 (1971): 76–136.

159. Niels van Doorn, "Platform Labor: On the Gendered and Racialized Exploitation of Low-Income Service Work in the 'On-Demand' Economy," *Information, Communication & Society* 20, no. 6 (2017): 898–914; Caroline O'Donovan, "If Most Of Your Income Comes from On-Demand Work, You're Probably a Racial Minority," *BuzzFeed News*, January 7, 2016, https://www.buzzfeednews.com/article/carolineodonovan/if-most-of-your-income-comes-from-on-demand-work-youre-proba; Bruce Reed, John Bridgel, and Conor McKay, "What Do On-Demand Economy Workers Want? We Asked Them," *HuffPost*, January 12, 2017, https://www.huffingtonpost.com/bruce-reed/what-do-ondemand-economy-_b_8958388.html.

160. Nunberg, "Goodbye Jobs, Hello 'Gigs.'"

3. Predict

1. Laura Santhanam and Vanessa Dennis, "What Do the Newly Released Witness Statements Tell Us about the Michael Brown Shooting?" *PBS NewsHour*, November 25, 2014, https://www.pbs.org/newshour/nation/newly-released-witness-testimony-tell-us-michael-brown-shooting.

2. James R. Martel, *Unburied Bodies: Subversive Corpses and the Authority of the Dead* (Amherst, Mass.: Amherst College Press, 2018).

3. Mark Follman, "Michael Brown's Mom Laid Flowers Where He Was Shot—and Police Crushed Them," *Mother Jones*, August 27, 2014, https://www.motherjones.com/politics/2014/08/ferguson-st-louis-police-tactics-dogs-michael-brown/.

4. Jon Swaine and Oliver Laughland, "Ferguson: More Arrests as Police and Protesters Clash for Second Night," *The Guardian*, August 11, 2015, https://www.theguardian.com/us-news/2015/aug/11/ferguson-protests-police-protesters-clash.

5. Yarimar Bonilla and Jonathan Rosa, "#Ferguson: Digital Protest, Hashtag Ethnography, and the Racial Politics of Social Media in the United States," *American Ethnologist* 42, no. 1 (2015): 4–17.

6. Gene Demby, "Does Having More Black Officers Reduce Police Violence?" *NPR*, February 4, 2017, https://www.npr.org/sections/codeswitch/2017/02/04/513

218656/does-having-more-black-officers-reduce-police-violence; Alex S. Vitale, *The End of Policing* (New York: Verso, 2017).

7. Vitale, *End of Policing.*

8. Data & Society Research Institute, "Zeke Edwards (ACLU) on Using Data to Make a Culture Change" (YouTube video, uploaded November 30, 2015), https://www.youtube.com/watch?time_continue=235&v=3lXJAL1riEc&feature=emb_logo.

9. Michelle Alexander, *The New Jim Crow: Mass Incarceration in the Age of Colorblindness* (New York: New Press, 2012); Douglas A. Blackmon, *Slavery by Another Name: The Re-Enslavement of Black Americans from the Civil War to World War II* (New York: Anchor Books, 2009); Edward E. Baptist, *The Half Has Never Been Told: Slavery and the Making of American Capitalism* (2014; repr., New York: Basic Books, 2016).

10. James Byrne and Gary Marx, "Technological Innovations in Crime Prevention and Policing: A Review of the Research on Implementation and Impact," *Journal of Police Studies* 3, no. 20 (2011): 17–40; James M. Byrne and Donald J. Rebovich, eds., *The New Technology of Crime, Law and Social Control* (Monsey, N.Y.: Criminal Justice Press, 2007); Christopher J. Harris, "The Police and Soft Technology: How Information Technology Contributes to Police Decision Making," in *The New Technology of Crime, Law and Social Control,* ed. James M. Byrne and Donald J. Rebovich (Monsey, N.Y.: Criminal Justice Press, 2007), 153–83.

11. Alex Pasternack, "The Big Business of Police Tech," *Fast Company,* June 19, 2017, https://www.fastcompany.com/40426359/the-big-business-of-police-tech.

12. Matt Stroud, "In a Week of Police Violence, Wall Street Won," *The Verge,* July 13, 2016, https://www.theverge.com/2016/7/13/12171756/police-violence-taser-motorola-northrup-firearm-stocks-jump.

13. Christine Byers, "Final Ferguson-Related Federal Report Dissects St. Louis County Police," *St. Louis Today,* October 2, 2015, https://www.stltoday.com/news/local/crime-and-courts/final-ferguson-related-federal-report-dissects-st-louis-county-police/article_33f1e567-1c22-57ec-8b6e-082197f05cf3.html; Christine Byers, "St. Louis County Police Chief Calls Justice Department Review a 'Missed Opportunity,'" *St. Louis Today,* January 2, 2017, https://www.stltoday.com/news/local/crime-and-courts/st-louis-county-police-chief-calls-justice-department-review-a/article_186d4d30-70ef-5afe-8efe-edd6f47bedf2.html.

14. Ryan J. Reilly and Christine Conetta, "State Senator to Ferguson Police: 'Will I Get Tear-Gassed Again?'" *Huffington Post,* August 13, 2014, https://www.huffingtonpost.com/2014/08/13/state-senator-ferguson_n_5676766.html.

15. Curtis Skinner, "Missouri State Senator Arrested by Ferguson Police," *Reuters,* October 21, 2014, https://www.reuters.com/article/us-usa-missouri-shooting-idUSKCN0IA0AM20141021.

16. Julia Angwin et al., "Machine Bias," *ProPublica,* May 23, 2016, https://www.propublica.org/article/machine-bias-risk-assessments-in-criminal-sentencing.

17. April Glaser, "Palantir Said It Had Nothing to Do with ICE Deportations. New Documents Seem to Tell a Different Story," *Slate,* May 3, 2019, https://slate.com/technology/2019/05/documents-reveal-palantir-software-is-used-for-ice-deportations.html.

18. Maurice Chammah, "Policing the Future," *The Verge,* 2016, https://www.theverge.com/2016/2/3/10895804/st-louis-police-hunchlab-predictive-policing-marshall-project.

19. David Robinson and Logan Koepke, *Stuck in a Pattern: Early Evidence on "Predictive Policing" and Civil Rights* (Washington, D.C.: Upturn, 2016), https://www.teamupturn.com/reports/2016/stuck-in-a-pattern.

20. Andy Cush, "Officer Go Fuck Yourself Is Out of a Job in Missouri," *Gawker,* August 29, 2014, http://gawker.com/officer-go-fuck-yourself-is-out-of-a-job-in-missouri-1628523901.

21. Chuck Wexler, *How Are Innovations in Technology Transforming Policing?* (Washington, D.C.: Police Executive Research Forum, 2012), https://www.policeforum.org/assets/docs/Critical_Issues_Series/how%20are%20innovations%20in%20technology%20transforming%20policing%202012.pdf.

22. Charlie Beck and Colleen McCue, "Predictive Policing: What Can We Learn from Wal-Mart and Amazon about Fighting Crime in a Recession?" *Police Chief* 76, no. 11 (2009): http://acmcst373ethics.weebly.com/uploads/2/9/6/2/29626713/police-chief-magazine.pdf.

23. Eric Siegel, "How to Fight Bias with Predictive Policing," *Scientific American: Voices* (blog), February 19, 2018, https://blogs.scientificamerican.com/voices/how-to-fight-bias-with-predictive-policing/.

24. Sam Corbett-Davies, Sharad Goel, and Sandra González-Bailon, "Even Imperfect Algorithms Can Improve the Criminal Justice System," *New York Times,* December 20, 2017, https://www.nytimes.com/2017/12/20/upshot/algorithms-bail-criminal-justice-system.html.

25. Rich Morin, Kim Parker, Renee Stepler, and Andrew Mercer, "Police Views, Public Views," *Pew Research Center: Social and Demographic Trends,* January 11, 2017, https://www.pewsocialtrends.org/2017/01/11/police-views-public-views/.

26. Solon Barocas and Andrew D. Selbst, "Big Data's Disparate Impact," *California Law Review* 104 (2016): 671–732; Andrew D. Selbst, "Disparate Impact in Predictive Policing," *Georgia Law Review* 52 (2017): 109–95.

27. Kate Crawford, "Artificial Intelligence's White Guy Problem," *New York Times,* June 26, 2016, https://www.nytimes.com/2016/06/26/opinion/sunday/artificial-intelligences-white-guy-problem.html; Steve Lohr, "Facial Recognition Is Accurate, If You're a White Guy," *New York Times,* February 9, 2018, https://www.nytimes.com/2018/02/09/technology/facial-recognition-race-artificial-intelligence.html; Sara Wachter-Boettcher, "How Silicon Valley's Blind Spots and Biases Are Ruining Tech for the Rest of Us," *Washington Post,* December 13, 2017,

https://www.washingtonpost.com/news/posteverything/wp/2017/12/13/how-silicon-valleys-blind-spots-and-biases-are-ruining-tech-for-the-rest-of-us/.

28. American Civil Liberties Union, "Predictive Policing Today: A Shared Statement of Civil Rights Concerns," August 31, 2016, https://www.aclu.org/other/statement-concern-about-predictive-policing-aclu-and-16-civil-rights-privacy-racial-justice.

29. Sarah Brayne, Alex Rosenblat, and danah boyd, "Predictive Policing: A Primer," *Data & Society,* October 27, 2015, https://datasociety.net/output/data-civil-rights-predictive-policing/; Danielle Ensign et al., "Runaway Feedback Loops in Predictive Policing," paper prepared for the first conference on Fairness, Accountability, and Transparency in Machine Learning, New York University, February 2018, http://arxiv.org/abs/1706.09847; Brian Jordan Jefferson, "Digitize and Punish: Computerized Crime Mapping and Racialized Carceral Power in Chicago," *Environment & Planning D: Society & Space* 35, no. 5 (2017): 775–96; Brian Jordan Jefferson, "Predictable Policing: Predictive Crime Mapping and Geographies of Policing and Race," *Annals of the American Association of Geographers* 108, no. 1 (2018): https://doi.org/10.1080/24694452.2017.1293500; Kristian Lum and William Isaac, "To Predict and Serve?," *Significance* 13, no. 5 (2016): 14–19.

30. Rashida Richardson, Jason Schultz, and Kate Crawford, "Dirty Data, Bad Predictions: How Civil Rights Violations Impact Police Data, Predictive Policing Systems, and Justice," *NYU Law Review* 94 (May 2019): 192–233.

31. Taylor Shelton, Matthew Zook, and Alan Wiig, "The 'Actually Existing Smart City,'" *Cambridge Journal of Regions, Economy and Society* 8, no. 1 (2015): 13–25.

32. Torin Monahan, "The Image of the Smart City: Surveillance Protocols and Social Inequality," in *Handbook of Cultural Security,* ed. Yasushi Watanabe (Cheltenham, UK: Edward Elgar, 2018), 210–26.

33. Kathleen Battles, *Calling All Cars: Radio Dragnets and the Technology of Policing* (Minneapolis: University of Minnesota Press, 2010); Joshua Reeves and Jeremy Packer, "Police Media: The Governance of Territory, Speed, and Communication," *Communication and Critical/Cultural Studies* 10, no. 4 (2013): 359–84.

34. Lucas D. Introna, "Algorithms, Governance, and Governmentality: On Governing Academic Writing," *Science, Technology, & Human Values* 41, no. 1 (2016): 20.

35. Heather Merrill and Lisa M. Hoffman, eds., *Spaces of Danger: Culture and Power in the Everyday* (Athens: University of Georgia Press, 2015); Loïc Wacquant, *Punishing the Poor: The Neoliberal Government of Social Insecurity* (Durham, N.C.: Duke University Press, 2009); Neil Websdale, *Policing the Poor: From Slave Plantation to Public Housing* (Boston: Northeastern University Press, 2001).

36. Battles, *Calling All Cars*; Vitale, *End of Policing*; Christopher J. Harris, "The Police and Soft Technology: How Information Technology Contributes to Police

Decision Making," in *The New Technology of Crime, Law and Social Control,* ed. James M. Byrne and Donald J. Rebovich (Monsey, N.Y.: Criminal Justice Press, 2007), 153–83; Dennis Rosenbaum, "Police Innovation Post-1980: Assessing Effectiveness and Equity Concerns in the Information Technology Era," *IPC Review* 1 (2007): 11–44; Peter K. Manning, *The Technology of Policing: Crime Mapping, Information Technology, and the Rationality of Crime Control* (New York: New York University Press, 2008); Christopher P. Wilson, *Cop Knowledge: Police Power and Cultural Narrative in Twentieth-Century America* (Chicago: University of Chicago Press, 2000).

37. Willem De Lint, "Autonomy, Regulation and the Police Beat," *Social & Legal Studies* 9, no. 1 (2000): 55–83; Willem De Lint, "Nineteenth Century Disciplinary Reform and the Prohibition against Talking Policemen," *Policing & Society* 9, no. 1 (1999): 33–58; Willem De Lint, "Arresting the Eye: Surveillance, Social Control and Resistance," *Space & Culture* 4, nos. 7–9 (2001): 21–49; Willem De Lint, "Regulating Autonomy: Police Discretion as a Problem for Training," *Canadian Journal of Criminology* 40, no. 3 (1998): 277–304.

38. Rachel Levinson-Waldman and Erica Posey, "Predictive Policing Goes to Court," *Brennan Center for Justice* (blog), September 5, 2017, https://www.brennancenter.org/blog/predictive-policing-goes-court; Rachel Levinson-Waldman and Erica Posey, "Court Rejects NYPD Attempts to Shield Predictive Policing from Disclosure," *Brennan Center for Justice* (blog), 2018, https://www.brennancenter.org/blog/court-rejects-nypd-attempts-shield-predictive-policing-disclosure; Ali Winston, "Transparency Advocates Win Release of NYPD 'Predictive Policing' Documents," *The Intercept,* January 27, 2018, https://theintercept.com/2018/01/27/nypd-predictive-policing-documents-lawsuit-crime-forecasting-brennan/.

39. Ali Winston, "Palantir Has Secretly Been Using New Orleans to Test Its Predictive Policing Technology," *The Verge,* February 27, 2018, https://www.theverge.com/2018/2/27/17054740/palantir-predictive-policing-tool-new-orleans-nopd; Darwin Bond-Graham, "All Tomorrow's Crimes: The Future of Policing Looks a Lot Like Good Branding," *San Francisco Weekly,* October 30, 2013, http://www.sfweekly.com/news/all-tomorrows-crimes-the-future-of-policing-looks-a-lot-like-good-branding/.

40. Lyndsey P. Beutin, "Racialization as a Way of Seeing: The Limits of Counter-Surveillance and Police Reform," *Surveillance & Society* 15, no. 1 (2017): 16.

41. Andrew Guthrie Ferguson, *The Rise of Big Data Policing: Surveillance, Race, and the Future of Law Enforcement* (New York: New York University Press, 2017), 29.

42. Jeff Asher and Rob Arthur, "Inside the Algorithm That Tries to Predict Gun Violence in Chicago," *New York Times,* June 13, 2017, https://www.nytimes.com/2017/06/13/upshot/what-an-algorithm-reveals-about-life-on-chicagos-high-risk-list.html.

43. Lynn Sweet and Fran Spielman, "Mayor Rahm Emanuel Fundraising in Washington, D.C., Thursday," *Chicago Sun-Times,* March 9, 2017, https://chicago.suntimes.com/2017/3/9/18373499/mayor-rahm-emanuel-fundraising-in-washington-d-c-thursday.

44. Ferguson, *Rise of Big Data Policing,* 29.

45. Other big data applications have also been considered for policing, including, for example, pilot studies to use algorithms to flag officers before they use excessive force. See Rob Arthur, "We Now Have Algorithms to Predict Police Misconduct," *FiveThirtyEight,* March 9, 2016, https://fivethirtyeight.com/features/we-now-have-algorithms-to-predict-police-misconduct/. However, police agencies appear far more likely to adopt predictive policing systems. A 2016 survey of the fifty largest police departments in the United States found that twenty agencies had already adopted predictive analytics for crime forecasting and eleven were actively considering it; see Robinson and Koepke, *Stuck in a Pattern.*

46. David Black, "Predictive Policing Is Here Now," *Manhattan Institute* (blog), February 29, 2016, https://www.manhattan-institute.org/html/predictive-policing-here-now-8563.html.

47. George L. Kelling and Mark H. Moore, "The Evolving Strategy of Policing," *Perspectives on Policing* 4 (1988): 1–16.

48. Vitale, *End of Policing.*

49. Simone Browne, *Dark Matters: On the Surveillance of Blackness* (Durham, N.C.: Duke University Press, 2015); Saidiya V. Hartman, *Scenes of Subjection: Terror, Slavery, and Self-Making in Nineteenth-Century America* (New York: Oxford University Press, 1997); Websdale, *Policing the Poor.*

50. Ronald F. Wright, "The Wickersham Commission and Local Control of Criminal Prosecution," *Marquette Law Review* 4 (2012): 1199–1220.

51. Jay Stuart Berman, *Police Administration and Progressive Reform: Theodore Roosevelt as Police Commissioner of New York* (Westport, Conn.: Greenwood Press, 1987); Gene E. Carte and Elaine H. Carte, *Police Reform in the United States: The Era of August Vollmer, 1905–1932* (Berkeley: University of California Press, 1975); David Alan Sklansky, "The Promise and Perils of Police Professionalism," in *The Future of Policing,* ed. Jennifer M. Brown (New York: Routledge, 2014), 343–54.

52. Kelling and Moore, "Evolving Strategy of Policing," 4.

53. David K. Johnson, *The Lavender Scare: The Cold War Persecution of Gays and Lesbians in the Federal Government* (Chicago: University of Chicago Press, 2006); Joy Rohde, "Police Militarization Is a Legacy of Cold War Paranoia," *The Conversation,* October 22, 2014, http://theconversation.com/police-militarization-is-a-legacy-of-cold-war-paranoia-32251.

54. Sacha Jenkins, dir., *Burn Motherfucker, Burn!* (New York: Showtime Documentary Films, 2017).

55. Khalil Gibran Muhammad, *The Condemnation of Blackness: Race, Crime, and the Making of Modern Urban America* (Cambridge, Mass.: Harvard University Press, 2011).

56. Beutin, "Racialization as a Way of Seeing," 16.

57. Kelling and Moore, "Evolving Strategy of Policing," 9.

58. Kelling and Moore.

59. David Garland, *The Culture of Control: Crime and Social Order in Contemporary Society* (Chicago: University of Chicago Press, 2001).

60. Garland; see also Nikolas Rose, "The Death of the Social? Re-Figuring the Territory of Government," *Economy & Society* 25, no. 3 (1996): 327–56.

61. Garland, *Culture of Control,* 16.

62. See also Michel Foucault, *The Birth of Biopolitics: Lectures at the Collège de France, 1978–1979,* ed. Michel Senellart, trans. Graham Burchell (New York: Palgrave Macmillan, 2008); Michaelis Lianos and Mary Douglas, "Dangerization and the End of Deviance: The Institutional Environment," *British Journal of Criminology* 40, no. 2 (2000): 261–78.

63. Kelling and Moore, "Evolving Strategy of Policing," 10.

64. Kelling and Moore, 8.

65. Bernard E. Harcourt, *Illusion of Order: The False Promise of Broken Windows Policing* (Cambridge, Mass.: Harvard University Press, 2001).

66. Kelling and Moore, "Evolving Strategy of Policing," 10.

67. Peter Miller and Nikolas Rose, *Governing the Present* (Malden, Mass.: Polity Press, 2008). See also John M. Dutton and William H. Starbuck, "Computer Simulation Models of Human Behavior: A History of an Intellectual Technology," *IEEE Transactions on Systems, Man, and Cybernetics* SMC-1, no. 2 (April 1971): 128–71.

68. Rosenbaum, "Police Innovation Post-1980;" Gregory F. Treverton et al., *Moving toward the Future of Policing* (Santa Monica, Calif.: RAND Corporation, 2011).

69. Treverton et al., *Moving toward the Future of Policing,* 33.

70. Manning, *Technology of Policing.*

71. Ingrid Burrington, "Policing Is an Information Business," *Urban Omnibus,* June 20, 2018, https://urbanomnibus.net/2018/06/policing-is-an-information-business/.

72. Burrington, "Policing Is an Information Business"; Ingrid Burrington, "The CompStat Evangelist Consultant World Tour," *Urban Omnibus,* June 20, 2018, https://urbanomnibus.net/2018/06/the-compstat-evangelist-consultant-world-tour/.

73. John A. Eterno and Eli B. Silverman, *The Crime Numbers Game: Management by Manipulation* (Boca Raton, Fla.: CRC Press, 2012).

74. George L. Kelling and James Q. Wilson, "Broken Windows: The Police and Neighborhood Safety," *Atlantic Monthly,* March 1982.

75. Harcourt, *Illusion of Order.*

76. Jeffrey Fagan et al., "Street Stops and Broken Windows Revisited: The Demography and Logic of Proactive Policing in a Safe and Changing City," in *Race, Ethnicity, and Policing. New and Essential Readings,* ed. Stephen K. Rice and Michael D. White (New York: New York University Press, 2009), 309–48; Bernard E. Harcourt, "Reflecting on the Subject: A Critique of the Social Influence Conception of Deterrence, the Broken Windows Theory, and Order-Maintenance Policing New York Style," *Michigan Law Review* 97, no. 2 (1998): 291–389; Bernard E. Harcourt and Jens Ludwig, "Broken Windows: New Evidence from New York City and a Five-City Social Experiment" (working paper, University of Chicago Law School, June 2005), http://papers.ssrn.com/abstract=743284; Babe Howell, "Broken Lives from Broken Windows: The Hidden Costs of Aggressive Order-Maintenance Policing," *New York University Review of Law & Social Change* 33 (2009): 271–329; Robert J. Sampson and Stephen W. Raudenbush, "Seeing Disorder: Neighborhood Stigma and the Social Construction of 'Broken Windows,'" *Social Psychology Quarterly* 67, no. 4 (2004): 319–42.

77. Marc Mauer, "Bill Clinton, 'Black Lives' and the Myths of the 1994 Crime Bill," *Marshall Project,* April 11, 2016, https://www.themarshallproject.org/2016/04/11/bill-clinton-black-lives-and-the-myths-of-the-1994-crime-bill.

78. Sentencing Project, "Fact Sheet: Trends in U.S. Corrections," n.d., https://sentencingproject.org/wp-content/uploads/2016/01/Trends-in-US-Corrections.pdf.

79. Black, "Predictive Policing Is Here Now."

80. Walter L. Perry et al., *Predictive Policing: The Role of Crime Forecasting in Law Enforcement Operations* (Santa Monica, Calif.: RAND Corporation, 2013), https://www.rand.org/content/dam/rand/pubs/research_reports/RR200/RR233/RAND_RR233.pdf.

81. Mark H. Moore and Anthony A. Braga, "Measuring and Improving Police Performance: The Lessons of Compstat and Its Progeny," *Policing: An International Journal* 26, no. 3 (2003): 439–53.

82. Beck and McCue, "Predictive Policing."

83. U.S. Department of Justice, "National Institute of Justice Predictive Policing Symposiums," November 18, 2009, and June 2–3, 2010, https://www.ncjrs.gov/pdffiles1/nij/242222and248891.pdf.

84. Stephen D. N. Graham, "Software-Sorted Geographies," *Progress in Human Geography* 29, no. 5 (2005): 562–80; Stephen Graham and David Wood, "Digitizing Surveillance: Categorization, Space, Inequality," *Critical Social Policy* 23, no. 2 (2003): 227–48.

85. CityCop, "Police Use Analytics to Reduce Crime (IBM Commercial)" (YouTube video, uploaded December 3, 2016), https://www.youtube.com/watch?v=5n2UjBO22EI.

86. William J. Bratton and Brian C. Anderson, "William Bratton on 'Precision Policing,'" *City Journal,* August 1, 2018, https://www.city-journal.org/html/william-bratton-precision-policing-16084.html.

87. Alan Binder, "Michael Slager, Officer in Walter Scott Shooting, Gets 20-Year Sentence," *New York Times,* December 7, 2017, https://www.nytimes.com/2017/12/07/us/michael-slager-sentence-walter-scott.html.

88. De Lint, "Autonomy, Regulation and the Police Beat," 55.

89. Markus D. Dubber and Mariana Valverde, eds., *The New Police Science: The Police Power in Domestic and International Governance* (Palo Alto, Calif.: Stanford University Press, 2006), 5; Markus D. Dubber and Mariana Valverde, eds., *Police and the Liberal State* (Palo Alto, Calif.: Stanford University Press, 2008).

90. De Lint, "Autonomy, Regulation and the Police Beat," 59.

91. Bratton and Anderson; "Sir Robert Peel's Nine Principles of Policing," *New York Times,* April 16, 2014, https://www.nytimes.com/2014/04/16/nyregion/sir-robert-peels-nine-principles-of-policing.html.

92. Cf. Dominic A. Wood, "The Importance of Liberal Values within Policing: Police and Crime Commissioners, Police Independence and the Spectre of Illiberal Democracy," *Policing and Society* 26, no. 2 (2016): 148–64.

93. Dubber and Valverde, *New Police Science,* 5.

94. De Lint, "Autonomy, Regulation and the Police Beat."

95. David Jones, *Crime, Protest, Community, and Police in Nineteenth-Century Britain* (London: Routledge, 1982).

96. Hartman, *Scenes of Subjection;* Amy Louise Wood, *Lynching and Spectacle: Witnessing Racial Violence in America, 1890–1940* (Chapel Hill: University of North Carolina Press, 2011).

97. Alexander, *The New Jim Crow;* Blackmon, *Slavery by Another Name;* Harcourt, *Illusion of Order.*

98. William Mazzarella, "Culture, Globalization, Mediation," *Annual Review of Anthropology* 33 (October 2004): 345–67; Sarah Kember and Joanna Zylinska, *Life after New Media: Mediation as a Vital Process* (Cambridge, Mass.: MIT Press, 2012).

99. Kember and Zylinska, *Life after New Media,* xv; see also Ian Hacking, "Making Up People," in *Reconstructing Individualism: Autonomy, Individuality, and the Self in Western Thought,* ed. Thomas C. Heller, Morton Sosna, and David E. Wellberry (Stanford, Calif.: Stanford University Press, 1986), 222–36; John Law and John Urry, "Enacting the Social," *Economy and Society* 33, no. 3 (2004): 390–410.

100. Mazzarella, "Culture, Globalization, Mediation," 346.

101. Mazzarella, 346.

102. Michel Foucault, *Security, Territory, Population: Lectures at the College de France, 1977–78,* trans. Graham Burchell (New York: Palgrave Macmillan, 2007); Michel Foucault, *Discipline and Punish: The Birth of the Prison,* trans. Alan Sheridan (New York: Penguin, 1977); Michel Foucault, *Power/Knowledge: Selected Interviews and Other Writings, 1972–1977,* ed. Colin Gordon, (New York: Vintage, 1980).

103. De Lint, "Autonomy, Regulation and the Police Beat," 64.
104. Reeves and Packer, "Police Media," 363.
105. Merrill and Hoffman, *Spaces of Danger.*
106. Jacques Rancière, *Dissensus: On Politics and Aesthetics,* trans. Steven Corcoran (New York: Continuum, 2010), 36. Rancière's notion of police exceeds the narrow definition of law enforcement, but nonetheless articulates the ambiguities of liberal consent policing under consideration; see Thomas Nail, *Theory of the Border* (New York: Oxford University Press, 2015), 116.
107. Mazzarella, "Culture, Globalization, Mediation," 357.
108. Garland, *Culture of Control;* Harcourt, *Illusion of Order.*
109. Nail, *Theory of the Border.*
110. Nail, 121.
111. Edwin Chadwick, quoted in Nail, 122.
112. Nail, 359–360.
113. Reeves and Packer, "Police Media," 378.
114. De Lint, "Autonomy, Regulation and the Police Beat."
115. De Lint, 70, emphasis added.
116. Lawrence W. Sherman, "The Rise of Evidence-Based Policing: Targeting, Testing, and Tracking," *Crime & Justice* 42, no. 1 (2013): 377–451.
117. Ingrid Burrington, "What Amazon Taught the Cops," *The Nation,* May 27, 2015, https://www.thenation.com/article/archive/what-amazon-taught-cops/.
118. Sherman, "Rise of Evidence-Based Policing," 379; see also Ferguson, *Rise of Big Data Policing.*
119. One exception is Bilel Benbouzid, "To Predict and to Manage: Predictive Policing in the United States," *Big Data & Society* 6, no. 1 (January 2019), https://doi.org/10.1177/2053951719861703.
120. Burrington, "What Amazon Taught the Cops."
121. "Azavea: Beyond Dots on a Map," n.d., accessed June 27, 2019, https://www.azavea.com/.
122. Robinson and Koepke, *Stuck in a Pattern;* Winston, "Palantir Has Secretly Been Using New Orleans."
123. For example, HunchLab was approached by a representative for a company that sells automated license-plate readers (ALPR), which was hoping for some sort of partnership arrangement. The team made a distinction between what they saw as predictive policing's unobtrusive forecasting and the intensive surveillance that ALPRs represented. Field notes, November 11, 2015.
124. Several other tools were developed using grant money from the National Institute of Justice; see National Institute of Justice, "Predictive Policing Research," January 13, 2014, https://www.nij.gov:443/topics/law-enforcement/strategies/predictive-policing/Pages/research.aspx.
125. In an interesting connection within the small world of predictive policing, Cheetham worked at the PPD under Commissioner John Timoney, who was

imported by former Philadelphia mayor Ed Rendell at the behest of William Bratton after a consultation on implementing Compstat; cf. Burrington, "Policing Is an Information Business."

126. Ralph B. Taylor, Jerry H. Ratcliffe, and Amber Perenzin, "Can We Predict Long-Term Community Crime Problems? The Estimation of Ecological Continuity to Model Risk Heterogeneity," *Journal of Research in Crime & Delinquency* 52, no. 5 (2015): 635–57.

127. Azavea, "HunchLab Predictive Missions at Greensboro PD: 'Tell me what I don't know!'" (YouTube video, uploaded November 25, 2015), https://www.youtube.com/watch?v=E-QdYqZzQhY.

128. Michael Townsley, Ross Homel, and Janet Chaseling, "Repeat Burglary Victimisation: Spatial and Temporal Patterns," *Australian & New Zealand Journal of Criminology* 33, no. 1 (2000): 37–63; Michael Townsley, Ross Homel, and Janet Chaseling, "Infectious Burglaries: A Test of the Near Repeat Hypothesis," *British Journal of Criminology* 43, no. 3 (2003): 615–33. The competing system PredPol uses a parsimonious forecasting technique that approximates the "near repeat" method. See P. Jeffrey Brantingham, Matthew Valasik, and George O. Mohler, "Does Predictive Policing Lead to Biased Arrests? Results from a Randomized Controlled Trial," *Statistics & Public Policy* 5, no. 1 (2018): 1–6; Brayne, Rosenblat, and boyd, "Predictive Policing"; George O. Mohler et al., "Randomized Controlled Field Trials of Predictive Policing," *Journal of the American Statistical Association* 110, no. 512 (2015): 1399–1411.

129. Joel M. Caplan, Leslie W. Kennedy, and Joel Miller, "Risk Terrain Modeling: Brokering Criminological Theory and GIS Methods for Crime Forecasting," *Justice Quarterly* 28, no. 2 (2011): 360–81.

130. For a pricing perspective, HunchLab's competitor PredPol charges about $200,000 per year. See Chammah, "Policing the Future."

131. Field notes, St. Louis County, Missouri, December 2, 2015. HunchLab tested the accuracy of its St. Louis County Police Department model to be included in a presentation to departmental command staff during a site visit.

132. Zack Quaintance, "What's New in Civic Tech: Top 10 Most Popular Types of Open Data," *Government Technology*, September 14, 2017, https://www.govtech.com/civic/Whats-New-in-Civic-Tech-Top-10-Most-Popular-Types-of-Open-Data.html.

133. Mohler et al., "Randomized Controlled Field Trials."

134. Karen Barad, "Posthumanist Performativity: Toward an Understanding of How Matter Comes to Matter," *Signs* 28, no. 3 (2003): 801–31.

135. Field notes, conversation with Jeremy Heffner, HunchLab offices. October 1, 2015.

136. Field notes, October 1, 2015.

137. Adrian Mackenzie, "The Performativity of Code Software and Cultures of Circulation," *Theory, Culture & Society* 22, no. 1 (2005): 71–92; Adrian Mackenzie,

"The Production of Prediction: What Does Machine Learning Want?" *European Journal of Cultural Studies* 18, nos. 4–5 (2015): 429–45.

138. Mackenzie, "Production of Prediction," 442.

139. Donna Haraway, "Situated Knowledges: The Science Question in Feminism and the Privilege of Partial Perspective," *Feminist Studies* 14, no. 3 (1988): 575.

140. Mackenzie, 443.

141. Sir Robert Peel, "Principles of Law Enforcement," 1829, https://www.durham.police.uk/About-Us/Documents/Peels_Principles_Of_Law_Enforcement.pdf.

142. Priscillia Hunt, Jessica Saunders, and John S. Hollywood, *Evaluation of the Shreveport Predictive Policing Experiment* (Santa Monica, Calif.: RAND Corporation, 2014), https://www.rand.org/pubs/research_reports/RR531.html; Jessica Saunders, Priscillia Hunt, and John S. Hollywood, *Predictions Put Into Practice: A Quasi-Experimental Evaluation of Chicago's Predictive Policing Pilot* (Santa Monica, Calif.: RAND Corporation, 2016), https://www.rand.org/pubs/external_publications/EP67204.html; Jerry H. Ratcliffe, Ralph B. Taylor, and Amber P. Askey, "The Philadelphia Predictive Policing Experiment: Effectiveness of the Prediction Models: Overview," Temple University Center for Security and Crime Science, 2017, https://liberalarts.temple.edu/sites/liberalarts/files/Predictive%20Policing%20Experiment%203_Efficacy.pdf.

143. Azavea, "Beyond the Box: Towards Prescriptive Analysis in Policing" (YouTube video, uploaded September 29, 2014), https://youtu.be/NCXFDfQsYBE.

144. Sam Ransbotham, David Kron, and Kirk Prentice, "Minding the Analytics Gap," *MIT Sloan Management Review*, Spring 2015.

145. Azavea, "Beyond the Box."

146. Field notes, HunchLab offices. May 5, 2016.

147. See Ratcliffe, Taylor, and Askey, "Philadelphia Predictive Policing Experiment."

148. Azavea, "Beyond the Box."

149. Cf. Benbouzid, "To Predict and to Manage."

150. Heffner routinely lamented that commanders placed too much faith in the predictions as a be-all and end-all, for instance, by requiring that officers spend lengthy spells in a predicted crime location despite the recommended "dosing" of ten to fifteen minutes in each cell. This is corroborated by extensive research on police department's adoption of predictive policing systems; see Sarah Brayne, "Big Data Surveillance: The Case of Policing," *American Sociological Review* 82, no. 5 (2017): 977–1008.

151. Amanda Geller et al., "Aggressive Policing and the Mental Health of Young Urban Men," *American Journal of Public Health* 104, no. 12 (2014): 2321–27; Fagan et al., "Street Stops and Broken Windows Revisited."

152. The Allocation Engine involves a set of mathematical rules dictating the selection of mission areas. These rules can be tweaked by clients to prioritize

strategic goals. Risk forecasts are transformed into z-scores, which are then used to filter out cells below a certain threshold, eliminating low-risk cells from being allocated as a Predictive Mission. Within the filtered collection of cells, weights are then used to differentiate between medium- and high-risk locations, and randomization is introduced in the selection within this narrowed set.

153. Azavea, "Beyond the Box."

154. Benbouzid, "To Predict and to Manage."

155. Kate J. Bowers et al., "Spatial Displacement and Diffusion of Benefits among Geographically Focused Policing Initiatives: A Meta-Analytical Review," *Journal of Experimental Criminology* 7, no. 4 (2011): 347–74; Anthony A. Braga, "Hot Spots Policing and Crime Prevention: A Systematic Review of Randomized Controlled Trials," *Journal of Experimental Criminology* 1, no. 3 (2005): 317–42; Rob T. Guerette and Kate J. Bowers, "Assessing the Extent of Crime Displacement and Diffusion of Benefits: A Review of Situational Crime Prevention Evaluations," *Criminology* 47, no. 4 (2009): 1331–68; Dennis P. Rosenbaum, "The Limits of Hot Spot Policing," in *Police Innovation: Contrasting Perspectives,* ed. David Weisburd and Anthony A. Braga (New York: Cambridge University Press, 2006), 245–63.

156. Emma Rey, "Predictive Policing and Crime Displacement," *PredPol* (blog), December 4, 2014, https://www.predpol.com/crime-displacement-predpol/; Martin B. Short et al., "Dissipation and Displacement of Hotspots in Reaction–Diffusion Models of Crime," *Proceedings of the National Academy of Sciences* 107, no. 9 (March 2010): 3961–65.

157. David Weisburd et al., "Does Crime Just Move around the Corner? A Controlled Study of Spatial Displacement and Diffusion of Crime Control Benefits," *Criminology* 44, no. 3 (2006): 549–92.

158. Scott W. Phillips, "Police Discretion and Boredom: What Officers Do When There Is Nothing to Do," *Journal of Contemporary Ethnography* 45, no. 5 (2015): 2.

159. Field notes, HunchLab offices, December 18, 2015.

160. Alene Tchekmedyian, "Police Push Back against Using Crime-Prediction Technology to Deploy Officers," *Los Angeles Times,* October 2, 2016, http://www.latimes.com/local/lanow/la-me-police-predict-crime-20161002-snap-story.html.

161. Dave Allen, quoted in Jane Wakefield, "Future Policing Will Go Hi-Tech," *BBC News,* July 3, 2013, https://www.bbc.com/news/technology-22954783.

162. Brayne, "Big Data Surveillance."

163. Ratcliffe has worked closely with the HunchLab team. In addition to introducing Robert Cheetham to the near-repeat pattern, he has also been hired as a consultant to advise on HunchLab's development.

164. Jerry Ratcliffe, personal communication, September 28, 2016.

165. Field notes, HunchLab offices, December 18, 2015.

166. Field notes, HunchLab offices, December 18, 2015.

167. Field notes, interview with Jeremy Heffner, March 15, 2015.

168. Field notes, December 2, 2015.

169. Azavea, "HunchLab Predictive Missions at Greensboro PD."

170. NYU School of Law, "Uneasy Partnerships: Private Industry and the Public Trust" (YouTube video, uploaded September 20, 2016,) https://www.youtube.com/watch?v=M1saeirVqqU.

171. NYU School of Law. Emphasis reflects speech.

172. Sorelle A. Friedler et al., "A Comparative Study of Fairness-Enhancing Interventions in Machine Learning," paper prepared for the Fairness, Accountability and Transparency in Machine Learning Conference, Halifax, NS, 2018, http://arxiv.org/abs/1802.04422; Muhammad Bilal Zafar et al., "Fairness Constraints: Mechanisms for Fair Classification," *Proceedings of the 20th International Conference on Artificial Intelligence and Statistics* 54 (2017), http://arxiv.org/abs/1507.05259.

173. Oded Berger-Tal et al., "The Exploration–Exploitation Dilemma: A Multidisciplinary Framework," *PLoS ONE* 9, no. 4 (2014): e95693.

174. See also Benbouzid, "To Predict and to Manage."

175. Bernard E. Harcourt, *Against Prediction: Profiling, Policing, and Punishing in an Actuarial Age* (Chicago: University of Chicago Press, 2006); Ensign et al., "Runaway Feedback Loops."

176. See Benbouzid, "To Predict and to Manage."

177. Muhammad, *The Condemnation of Blackness.*

178. Randomized control trials have been the method of choice especially since the 1995 publication of Lawrence Sherman and David Weisburd's Minneapolis "hot spots" experiment; see Lawrence W. Sherman and David Weisburd, "General Deterrent Effects of Police Patrol in Crime 'Hot Spots': A Randomized, Controlled Trial," *Justice Quarterly* 12, no. 4 (1995): 625–48; Cynthia Lum and Lorraine Mazerolle, "History of Randomized Controlled Experiments in Criminal Justice," in *Encyclopedia of Criminology and Criminal Justice,* ed. Gerben Bruinsma and David Weisburd (New York: Springer, 2014), 2227–39.

179. Azavea, "HunchLab Advisor: Know What Works" (YouTube video, uploaded February 5, 2015), https://www.youtube.com/watch?v=hHDJfHPYTsU.

180. Azavea.

181. The challenge was canceled as of January 2016 because no departments applied for funding. Gregory Ridgeway, personal communication, January 17, 2018.

182. Debra Cohen McCullough and Deborah L. Spence, eds., *American Policing in 2022: Essays on the Future of a Profession* (Washington, D.C.: Community Oriented Policing Services, National Institute of Justice, 2012).

183. Sherman, "Rise of Evidence-Based Policing."

184. Azavea, "HunchLab Advisor."

185. Azavea.

186. Karl Popper, *The Logic of Scientific Discovery* (London: Routledge, 1959).

187. Chris Anderson, "The End of Theory: The Data Deluge Makes the Scientific Method Obsolete," *Wired,* June 2008, https://www.wired.com/2008/06/pb-theory/; Viktor Mayer-Schönberger and Kenneth Cukier, *Big Data* (Boston: Houghton Mifflin, 2012). See also Mark Andrejevic and Kelly Gates, "Big Data Surveillance: Introduction," *Surveillance & Society* 12, no. 2 (2014): 185–96.

188. Brian Jordan Jefferson, "Predictable Policing: Predictive Crime Mapping and Geographies of Policing and Race," *Annals of the American Association of Geographers* 108, no. 1 (2017): 1–16; Robinson and Koepke, *Stuck in a Pattern.*

189. Blake Norton et al., *Collaborative Reform Initiative: An Assessment of the St. Louis County Police Department* (Washington, D.C.: Police Foundation, 2015), https://www.policefoundation.org/publication/collaborative-reform-initiative-an-assessment-of-the-st-louis-county-police-department/.

190. Summer Ballentine, "Black Missouri Drivers 91% More Likely to Be Stopped, State Attorney General Finds," *PBS NewsHour,* June 10, 2019, https://www.pbs.org/newshour/nation/black-missouri-drivers-91-more-likely-to-be-stopped-state-attorney-general-finds.

191. Andrew Guthrie Ferguson, "The Exclusionary Rule in the Age of Blue Data," *Vanderbilt Law Review* 72 (2019): 561; see also Ferguson, *Rise of Big Data Policing.*

192. Cf. Jordan Blair Woods, "Decriminalization, Police Authority, and Routine Traffic Stops," *UCLA Law Review* 3 (2015): 672–759.

193. Beck and McCue, "Predictive Policing."

194. Siegel, "How to Fight Bias with Predictive Policing."

195. Brian Massumi, "Potential Politics and the Primacy of Preemption," *Theory & Event* 10, no. 2 (2007), https://www.doi.org/10.1353/tae.2007.0066.

196. Wendy Hui Kyong Chun, "Ways of Knowing (Cities) Networks," paper prepared for Ways of Knowing Cities Symposium, Center for Spatial Research, Graduate School of Architecture, Planning and Preservation, Columbia University, February 9, 2018, https://www.arch.columbia.edu/events/816-ways-of-knowing-cities; Elvia Wilk, "The Proxy and Its Politics: On Evasive Objects in a Networked Age," *Rhizome,* July 25, 2017, http://rhizome.org/editorial/2017/jul/25/the-proxy-and-its-politics/.

197. An exception is Brayne, Rosenblat and boyd's primer on predictive policing, titled "Predictive Policing," which raises questions about the relationship between algorithmic predictions and current best practices in police patrol.

198. Dubber and Valverde, *New Police Science.*

199. Jefferson, "Digitize and Punish," 778, emphasis added.

200. Introna, "Algorithms, Governance, and Governmentality;" Jefferson, "Digitize and Punish"; Jefferson, "Predictable Policing."

201. Robert Cheetham, "Why We Sold HunchLab," *Azavea* (blog), January 23, 2019, https://www.azavea.com/blog/2019/01/23/why-we-sold-hunchlab/.

202. Cheetham.

203. De Lint, "Autonomy, Regulation and the Police Beat," 70.

204. John Locke, *Two Treatises of Government*, ed. Peter Laslett (New York: Cambridge University Press, 1988).

Conclusion

1. Kevin Rogan, "The 3 Pictures That Explain Everything about Smart Cities," *CityLab*, June 27, 2019, https://www.citylab.com/design/2019/06/smart-city-photos-technology-marketing-branding-jibberjabber/592123/.

2. Jonathan Woetzel et al., *Smart Cities: Digital Solutions for a More Livable Future* (San Francisco: McKinsey Global Institute, 2018), https://www.mckinsey.com/industries/capital-projects-and-infrastructure/our-insights/smart-cities-digital-solutions-for-a-more-livable-future.

3. Orit Halpern et al., "Test-Bed Urbanism," *Public Culture* 25, no. 270 (2013): 272–306.

4. Daniel Markovits, "How McKinsey Destroyed the Middle Class," *The Atlantic*, February 3, 2020, https://www.theatlantic.com/ideas/archive/2020/02/how-mckinsey-destroyed-middle-class/605878/.

5. Sandro Mezzadra and Brett Neilson, *The Politics of Operations: Excavating Contemporary Capitalism* (Durham, NC: Duke University Press, 2019).

6. *Merriam Webster's Collegiate Dictionary*, 11th ed. (2003), s.v. "efficiency."

7. AbdouMaliq Simone, "The Surfacing of Urban Life," *City* 15, nos. 3–4 (2011): 356.

8. Thomas Osborne and Nikolas Rose, "Governing Cities: Notes on the Spatialization of Virtue," *Environment & Planning D* 17 (1999): 737–60.

9. Henri Lefebvre, *The Urban Revolution*, trans. Robert Bononno (Minneapolis: University of Minnesota Press, 2003); Andy Merrifield, *The New Urban Question* (London: Pluto Press, 2014).

10. Ryan Bishop and John W. P. Phillips, "The Urban Problematic," *Theory, Culture & Society* 30, nos. 7–8 (2013): 223.

11. Karl Polanyi, *The Great Transformation: The Political and Economic Origins of Our Time*, 2nd ed. (Boston: Beacon Press, 2001).

12. Gilles Deleuze, *Foucault*, trans. Seán Hand (Minneapolis: University of Minnesota Press, 1988), 44.

13. Brett Neilson, "Five Theses on Understanding Logistics as Power," *Distinktion: Journal of Social Theory* 13, no. 3 (2012): 322–39.

14. Kate Hepworth, "Enacting Logistical Geographies," *Environment and Planning D: Society and Space* 32, no. 6 (2014): 1121.

15. Richard Florida, "The Rise of 'Urban Tech,'" *CityLab*, July 10, 2018, https://www.citylab.com/life/2018/07/the-rise-of-urban-tech/564653/.

16. Woetzel et al., *Smart Cities*.

17. Eshwar Nurani-Parasuraman, "So, What Is the Smartness Quotient of Your City?" *Schneider Electric Blog,* August 18, 2013, https://blog.se.com/smart-grid/2013/08/18/so-what-is-the-smartness-quotient-of-your-city/.

18. Frank Pasquale, *The Black Box Society: The Secret Algorithms That Control Money and Information* (2015; repr., Cambridge, Mass.: Harvard University Press, 2016).

19. David Hill, "The Injuries of Platform Logistics," *Media, Culture & Society* (2019), forthcoming, https://ray.yorksj.ac.uk/id/eprint/3887/3/Platform%20Logistics.pdf.

20. Ava Kofman, "Are New York's Free LinkNYC Internet Kiosks Tracking Your Movements?" *The Intercept,* September 8, 2018, https://theintercept.com/2018/09/08/linknyc-free-wifi-kiosks/; Nick Pinto, "Google Is Transforming NYC's Payphones Into a 'Personalized Propaganda Engine,'" *Village Voice,* July 6, 2016, https://www.villagevoice.com/2016/07/06/google-is-transforming-nycs-payphones-into-a-personalized-propaganda-engine/.

21. Michel Foucault, *Discipline and Punish: The Birth of the Prison,* trans. Alan Sheridan (New York: Penguin, 1977).

22. Jathan Sadowski and Frank Pasquale, "The Spectrum of Control: A Social Theory of the Smart City," *First Monday* 20, no. 7 (2015), https://doi.org/10.5210/fm.v20i7.5903.

23. Gilles Deleuze, "Postscript on the Societies of Control," *October* 59 (Winter 1992): 4. See also Arjun Appadurai, "The Wealth of Dividuals," in *Derivatives and the Wealth of Societies,* ed. Benjamin Lee and Randy Martin (Chicago: University of Chicago Press, 2016), 17–36.

24. Deleuze, "Postscript"; John Cheney-Lippold, "A New Algorithmic Identity: Soft Biopolitics and the Modulation of Control," *Theory, Culture & Society* 28, no. 6 (2011): 169.

25. Jennifer L. Croissant, "Agnotology: Ignorance and Absence or Towards a Sociology of Things That Aren't There," *Social Epistemology* 28, no. 1 (2014): 4–25; Linsey McGoey, "Strategic Unknowns: Towards a Sociology of Ignorance," *Economy and Society* 41, no. 1 (2012): 1–16.

26. Ganaele Langlois and Greg Elmer, "Impersonal Subjectivation from Platforms to Infrastructures," *Media, Culture & Society* 41, no. 2 (March 2019): 238; Adrian Mackenzie, "Personalization and Probabilities: Impersonal Propensities in Online Grocery Shopping," *Big Data & Society* (January–June 2018), https://doi.org/10.1177/2053951718778310.

27. Brian Massumi, *99 Theses on the Revaluation of Value: A Postcapitalist Manifesto* (Minneapolis: University of Minnesota Press, 2018).

28. John Cheney-Lippold, *We Are Data: Algorithms and the Making of Our Digital Selves* (New York: New York University Press, 2017).

29. Jeremy Packer, "Epistemology Not Ideology, or Why We Need New Germans," *Communication and Critical/Cultural Studies* 10, nos. 2–3 (2013): 298.

30. Michaelis Lianos and Mary Douglas, "Dangerization and the End of Deviance: The Institutional Environment," *British Journal of Criminology* 40, no. 2 (2000): 261–78; see also Martin Dodge and Rob Kitchin, *Code/Space: Software and Everyday Life* (Cambridge, Mass.: MIT Press, 2011).

31. Lianos and Douglas, "Dangerization and the End of Deviance," 264–65.

32. Deleuze, "Postscript," 7.

33. Benjamin H. Bratton, *The Stack: On Software and Sovereignty* (Cambridge, Mass.: MIT Press, 2016); Alexander R. Galloway, *Protocol: How Control Exists after Decentralization* (Cambridge, Mass.: MIT Press, 2004).

34. Giorgio Agamben, *Homo Sacer: Sovereign Power and Bare Life,* trans. Daniel Heller-Roazen (Stanford, Calif.: Stanford University Press, 1998).

35. Deleuze, *Foucault,* 43.

36. Jasper Bernes, "Logistics, Counterlogistics and the Communist Prospect," *Endnotes* 3 (2013), https://endnotes.org.uk/issues/3/en/jasper-bernes-logistics-counterlogistics-and-the-communist-prospect.

37. Bernes; Alberto Toscano, "Logistics and Opposition," *Mute,* August 9, 2011, http://www.mctamute.org/editorial/articles/logistics-and-opposition; Alberto Toscano, "Lineaments of the Logistical State," *Viewpoint,* September 29, 2014, https://www.viewpointmag.com/2014/09/28/lineaments-of-the-logistical-state/.

38. Marco Briziarelli, "Spatial Politics in the Digital Realm: The Logistics/Precarity Dialectics and Deliveroo's Tertiary Space Struggles," *Cultural Studies,* September 19, 2018, 1–18.

39. Darwin Bond-Graham, "All Tomorrow's Crimes: The Future of Policing Looks a Lot Like Good Branding," *SF Weekly,* October 30, 2013, http://www.sfweekly.com/news/all-tomorrows-crimes-the-future-of-policing-looks-a-lot-like-good-branding/. Rachel Levinson-Waldman and Erica Posey, "Predictive Policing Goes to Court," *Brennan Center for Justice,* September 5, 2017, https://www.brennancenter.org/blog/predictive-policing-goes-court; Ali Winston, "Palantir Has Secretly Been Using New Orleans to Test Its Predictive Policing Technology," *The Verge,* February 27, 2018, https://www.theverge.com/2018/2/27/17054740/palantir-predictive-policing-tool-new-orleans-nopd.

40. Brett Neilson, "The Reverse of Engineering," *South Atlantic Quarterly* 119, no. 1 (2020): 75–93.

41. Jonathan Blitzer, "A Veteran ICE Agent, Disillusioned with the Trump Era, Speaks Out," *New Yorker,* July 24, 2017, http://www.newyorker.com/news/news-desk/a-veteran-ice-agent-disillusioned-with-the-trump-era-speaks-out.

42. Obama claimed that his administration would invest in deporting "felons not families"—immigrants with violent criminal histories. However, investigative journalists have shown this to be false. Reporters at the *Marshall Project* discovered that about 60 percent of deportations in the last two years of the Obama administration targeted immigrants "with no criminal conviction or whose only crime was immigration-related, such as illegal entry or re-entry." See Christie

Thompson and Anna Flagg, "Who Is ICE Deporting?" *Marshall Project,* September 26, 2016, https://www.themarshallproject.org/2016/09/26/who-is-ice-deporting. However, the rate of deportations under President Trump's term has still been higher on average than under Obama. See Anna O. Law, "This Is How Trump's Deportations Differ from Obama's," *Washington Post,* May 3, 2017, https://www.washingtonpost.com/news/monkey-cage/wp/2017/05/03/this-is-how-trumps-deportations-differ-from-obamas/.

43. Alan Gomez, "Immigration Arrests up, Deportations down under Trump," *USA Today,* July 17, 2017, https://www.usatoday.com/story/news/world/2017/07/17/immigration-arrests-up-deportations-down-under-trump/484437001/; Jeff Sommer, "Trump Immigration Crackdown Is Great for Private Prison Stocks," *New York Times,* March 10, 2017, https://www.nytimes.com/2017/03/10/your-money/immigrants-prison-stocks.html.

44. Spencer Woodman, "Palantir Provides the Engine for Donald Trump's Deportation Machine," *The Intercept,* March 2, 2017, https://theintercept.com/2017/03/02/palantir-provides-the-engine-for-donald-trumps-deportation-machine/.

45. Jacques Peretti, "Palantir: The 'Special Ops' Tech Giant That Wields as Much Real-World Power as Google," *The Guardian,* July 30, 2017, http://www.theguardian.com/world/2017/jul/30/palantir-peter-thiel-cia-data-crime-police.

46. Woodman, "Palantir Provides the Engine."

47. Department of Homeland Security, "About Fusion Centers," July 6, 2009, https://www.dhs.gov/state-and-major-urban-area-fusion-centers; Aaron Shapiro, "The Future of Security Lies Beneath the City of Brotherly Love," *Annenberg Now* (blog), June 10, 2014, https://www.asc.upenn.edu/news-events/news/future-security-lies-beneath-city-brotherly-love.

48. Caroline Haskins, "Revealed: This Is Palantir's Top-Secret User Manual for Cops," *Vice,* July 12, 2019, https://www.vice.com/en_us/article/9kx4z8/revealed-this-is-palantirs-top-secret-user-manual-for-cops.

49. New Sanctuary Movement of Philadelphia, "Sanctuary in the Streets: Raid Response Guide," February 2017, http://sanctuaryphiladelphia.org/wp-content/uploads/2017/02/Sanctuary-in-the-Streets-Guide-Feb-2017.pdf.

50. Aubrey Whelan, "In West Philly, Activists Train to 'Disrupt' Deportations," Philly.com, January 28, 2017, http://www.philly.com/philly/news/In-West-Philly-activists-train-to-disrupt-deportations.html.

51. Michael E. Miller, "They Fear Being Deported, but 2.9 Million Immigrants Must Check in with ICE Anyway," *Washington Post,* April 25, 2019, https://www.washingtonpost.com/local/they-fear-being-deported-but-29-million-immigrants-must-check-in-with-ice-anyway/2019/04/25/ac74efce-6309-11e9-9ff2-abc984dc9eec_story.html.

52. Aaron Glickman and Janet Weiner, "Why Deaths Continue to Rise in the Opioid Epidemic," Leonard Davis Institute of Health Economics, University of

Pennsylvania, January 22, 2019, https://ldi.upenn.edu/healthpolicysense/why-deaths-continue-rise-opioid-epidemic.

53. Michelle Alexander, *The New Jim Crow: Mass Incarceration in the Age of Colorblindness* (New York: New Press, 2012); Deborah J. Vagins and Jesselyn McCurdy, *Cracks in the System: 20 Years of the Unjust Federal Crack Cocaine Law* (New York: American Civil Liberties Union, 2006), https://www.aclu.org/other/cracks-system-20-years-unjust-federal-crack-cocaine-law.

54. Donna Murch, "How Race Made the Opioid Crisis," *Boston Review,* April 9, 2019, http://bostonreview.net/forum/donna-murch-how-race-made-opioid-crisis.

55. See for example Roni Caryn Rabin, "New York Sues Sackler Family Members and Drug Distributors," *New York Times,* March 28, 2019, https://www.nytimes.com/2019/03/28/health/new-york-lawsuit-opioids-sacklers-distributors.html.

56. Harm Reduction Coalition, "Principles of Harm Reduction," n.d., https://harmreduction.org/about-us/principles-of-harm-reduction/.

57. John Hirst, "Meeting People Where They're At: Harm Reduction at the Red Door Clinic," *National Association of County and City Health Officials: Stories from the Field* (blog), March 29, 2016, https://www.nacchostories.org/meeting-people-where-theyre-at-harm-reduction-at-the-red-door-clinic/.

58. Annie Correal, "Once It Was Overdue Books: Now Librarians Fight Overdoses," *New York Times,* February 28, 2018, https://www.nytimes.com/2018/02/28/nyregion/librarians-opioid-heroin-overdoses.html.

59. Mike Newall, "For These Philly Librarians, Drug Tourists and Overdose Drills Are Part of the Job," *Philadelphia Inquirer,* May 21, 2017, https://www.inquirer.com/philly/columnists/mike_newall/opioid-crisis-Needle-Park-McPherson-narcan.html.

60. Correal, "Once It Was Overdue Books."

61. Christopher Moraff, "Over 200 Philly Heroin Users Poisoned over the Weekend, Potentially Due to 'K2,'" *Filter,* September 25, 2018, https://filtermag.org/2018/09/25/over-200-philly-heroin-users-poisoned-over-the-weekend-potentially-due-to-synthetic-cannabinoid-contamination/.

62. Bad Batch Alert, n.d., www.badbatchalert.com/. See also Stephen Babcock, "Baltimore Teens Built a System That Sends Text Alerts during Heroin Overdose Spikes," *Technical.ly Baltimore,* July 6, 2017, https://technical.ly/baltimore/2017/07/06/bad-batch-alert-legrand/.

63. City of Philadelphia, *SmartCityPHL Roadmap* (Philadelphia: Office of Innovation & Technology, 2019), https://www.phila.gov/media/20190204121858/SmartCityPHL-Roadmap.pdf.

64. City of Philadelphia.

65. Ryan Johnston, "Philadelphia Releases 'Smart City' Roadmap with Eyes on Tech Pitfalls," *StateScoop,* February 6, 2019, https://statescoop.com/philadelphia-releases-smart-city-roadmap-with-eyes-on-tech-pitfalls/.

66. Thomas Ginsberg, *Philadelphia's Immigrants: Who They Are and How They Are Changing the City* (Philadelphia: Pew Charitable Trusts, June 2018), https://www.pewtrusts.org/-/media/assets/2018/06/pri_philadelphias_immigrants.pdf.

67. City of Philadelphia, *SmartCityPHL Roadmap*.

68. Steven J. Jackson, "Rethinking Repair," in *Media Technologies: Essays on Communication, Materiality, and Society,* ed. Tarleton Gillespie, Pablo J. Boczkowski, and Kirsten A. Foot (Cambridge, Mass.: MIT Press, 2014), 227.

69. Sung-Yueh Perng and Sophia Maalsen, "Civic Infrastructure and the Appropriation of the Corporate Smart City," *Annals of the American Association of Geographers*, November 11, 2019, 2, emphasis added.

Index

AARON SHAPIRO is assistant professor of technology studies in the Department of Communication at the University of North Carolina, Chapel Hill.